Poverty and Development

Full details of all of our publications can be found on http://www.chan nelviewpublications.com, or by writing to Channel View Publications, St Nicholas House, 31–34 High Street, Bristol BS1 2AW, UK.

Poverty and Development

Problems and Prospects

Edited by
Michal Apollo and Pahlaj Moolio

CHANNEL VIEW PUBLICATIONS
Bristol • Jackson

DOI https://doi.org/10.21832/APOLLO8472
Library of Congress Cataloging in Publication Data
A catalog record for this book is available from the Library of Congress.
Names: Apollo, Michal, 1981- editor. | Moolio, Pahlaj, editor.
Title: Poverty and Development: Problems and Prospects/Edited by Michal Apollo and Pahlaj Moolio.
Description: Blue Ridge Summit: Channel View Publications, [2022] | Includes bibliographical references and index. | Summary: "This book brings together interdisciplinary perspectives to broaden understandings of poverty. It contains empirical and conceptual chapters on a range of topics highlighting the relationship between poverty and sustainability. It will be a helpful resource for students and researchers of sustainable and regional development, and tourism"— Provided by publisher. Identifiers: LCCN 2021025836 (print) | LCCN 2021025837 (ebook) | ISBN 9781845418465 (pbk) | ISBN 9781845418472 (hbk) | ISBN 9781845418489 (pdf) | ISBN 9781845418496 (epub)
Subjects: LCSH: Poverty. | Sustainability. | Rural development. | Economic development. Classification: LCC HC79.P6 P68265 2022 (print) | LCC HC79.P6 (ebook) | DDC 339.4/6—dc23 LC record available at https://lccn.loc.gov/2021025836
LC ebook record available at https://lccn.loc.gov/2021025837

British Library Cataloguing in Publication Data
A catalogue entry for this book is available from the British Library.

ISBN-13: 978-1-84541-847-2 (hbk)
ISBN-13: 978-1-84541-846-5 (pbk)

Channel View Publications
UK: St Nicholas House, 31–34 High Street, Bristol, BS1 2AW, UK.
USA: Ingram, Jackson, TN, USA.

Website: www.channelviewpublications.com
Twitter: Channel_View
Facebook: https://www.facebook.com/channelviewpublications
Blog: www.channelviewpublications.wordpress.com

Typeset by Nova Techset Private Limited, Bengaluru and Chennai, India.

Wasteland

running running
*
lepers in square 155
*
wasteland in my arms
wasteland in my heart
wasteland all I see
*
running running
*
lepers in square 155
*
wasteland in my mouth
wasteland in my eyes
wasteland all I see
*
running running
*
lepers in square 155
*
wasteland in my mind
wasteland in my mouth
wasteland all I see

'*Wasteland*' comes from the album '*Headache*' performed by Trupa Trupa band
Text by Grzegorz Kwiatkowski and Wojciech Juchniewicz

Contents

Contributors

Editors

Michal Apollo is an Assistant Professor in the Institute of Geography, Department of Tourism and Regional Studies at the Pedagogical University of Cracow, Poland, a Fellow of Yale University's Global Justice Program, New Haven, USA, a Visiting Scholar at Hainan University – Arizona State University Joint International Tourism College, Haikou, China and a Visiting Fellow at the Center for Tourism Research, Wakayama University, Japan. He received his MSc and PhD in earth science in the field of geography at the Pedagogical University of Cracow, and a PGCert in global development from the University of Warsaw, Poland. Michal is an enthusiastic researcher (research field: man–environment systems), traveller (he has visited more than 60 countries on six continents), mountaineer (several new climbing routes, including the first ascent on two Himalaya peaks: in 2006, Masala Peak, and in 2012, Forgotten Peak), ultra-runner (100-miler finisher), diver (Advanced Open Water Diver), photographer (a few photo exhibitions) and science populariser (a few hundred pop-science lectures). Michal's unique background allows him to integrate knowledge from various perspectives into his research and consultancy work. His areas of expertise are tourism management, consumer behaviours and environmental and socioeconomic issues. In his main research field, he focuses on human presence and well-being in high-mountain regions. Currently, he is working on a concept for the sustainable use of environmental and human resources, as this is the key to the development, prosperity and well-being of all stakeholders. Michal is a founder and board member of the Polish Chapter of Academics Stand Against Poverty. ORCID: 0000-0002-7777-5176

Pahlaj Moolio is a Professor and Dean in the Office of Research and Development at LSi Business School, Phnom Penh, Cambodia, and a Fellow of Yale University's Global Justice Program, New Haven, USA. He studied mathematics, statistics and economics at the University of Sindh, Pakistan and the University of Chittagong, Bangladesh, completing his PhD in mathematical economics in 2011. After teaching in the Public School of Hyderabad, Pakistan, he lectured at the University for Education and Development and the University of Management and Economics in Battambang, and at the Paññāsāstra University of Cambodia in Phnom Penh, before joining the LSi Business School in 2021. Pahlaj was invited to serve as a DAAD (German Academic Exchange Program) visiting professor at the Ruhr-Universität Bochum, Germany (April–September 2019). His areas of research include official development aid, foreign direct investment, monetary policy, and migration in terms of rural development and poverty eradication, with an overarching interest in the building of social capital for sustainable development. His ongoing research interests in the field of labour migration earned him a Swedish International Development Agency (SIDA)-funded regional research project (2016–2017) under the SHAPE-SEA Research Program, which involved research in five countries in the ASEAN region. Currently, he is working on a study on narrowing the development gap in rural development and poverty eradication in Cambodia, Lao PDR, Myanmar and Vietnam, funded by the ASEAN-China Cooperation Fund. Pahlaj is the author/co-author of 16 research papers and two books. Moreover, he has served as a member and activist of the Minority's Rights Commission of Bangladesh (1997–2002) and as a member of the board of directors of the Santitham Education and Development Foundation in Thailand (2004–2008). He is also a founding member and co-chair of the Academics Stand Against Poverty (ASAP) Cambodian Chapter, an advisory board member at the Basel Institute of Commons and Economics in Switzerland, and a member of the ASAP Global Board of Directors. ORCID: 0000-0002-0769-9876

Authors

Mosope Akinbola is a graduate in sociology from the University of Lagos, Nigeria. He has co-authored and presented a number of academic papers even while he was an undergraduate at the University of Lagos, one of which includes a paper presented at the Second Annual Conference of Inclusivity, Equality and Diversity held in Lagos, Nigeria in 2019. Mosope is currently a registered member of the National Youth Service Corps (NYSC) in Nigeria. ORCID: 0000-0002-5519-918X

Muhammad Ashfaq is a Professor of Finance and Accounting at IU International University of Applied Sciences in Germany. He is also the Programme Director of Digital Business. He earned his PhD on the topic of 'Knowledge, attitude and practices toward Islamic banking and finance: An empirical analysis of retail consumers in Germany' from Tübingen University. Muhammad holds an MBA in financial management from Coburg University. He is a regular senior visiting faculty member at Wittenborg University of Applied Sciences in the Netherlands and the international MBA program at Coburg University. His research interests include Islamic finance and FinTech.

Celina Aznarez is a junior researcher at the Basque Centre for Climate Change, and a PhD student at the Institut de Ciència i Tecnologia

Ambientals (ICTA-UAB). Celina's areas of interest are ecosystem services modelling, as well as environmental justice and socio-ecological and distributional issues. Currently, she is working on the effects of rewilding strategies on the well-being of urban communities through the lens of ecosystem services research. ORCID: 0000-0001-7807-1413

Kennedy Eborka (PhD) is a lecturer in the Department of Sociology at the University of Lagos, Nigeria. His research interests span population and development studies, migration and urban poverty. He is a member of various professional organisations such as the International Union for the Scientific Study of Population (IUSSP), Planned Parenthood Federation of Nigeria (PPFN), the Nigerian Anthropological and Sociological Association (NASA), etc. Kennedy is an associate member of the Institute of African and Diaspora Studies (IADS), University of Lagos, and is currently collaborating with Academics Stand Against Poverty (ASAP), a global body of academics working to reduce poverty around the world. ORCID: 0000-0002-8652-1450

Agnieszka Gawlik is an Assistant Professor at WSB University in Opole, Poland. Agnieszka conducts research on the effectiveness of markets and market failures (e.g. tourism market, healthcare market). She is the author of several dozen publications on, among others, market and organisational effectiveness, innovations and competitiveness. Her research combines business management methods and techniques with specialist knowledge in the field of economics and finance. Agnieszka is also a passionate lecturer, advisor and consultant who combines an economic approach with care for the natural environment and social development. ORCID: 0000-0001-5647-8896

Cristina Ghinea is a lecturer in the Faculty of Food Engineering at Stefan cel Mare University of Suceava, Romania. She holds a PhD in environmental engineering (2012), co-supervised between TUIASI, Romania and the University of Twente, The Netherlands. Cristina's research interests are consumer behaviour, waste generation, food waste, environmental protection, waste management and processes for waste valorisation, statistical analysis, lifecycle assessment, waste management modelling and optimisation. ORCID:0000-0002-7744-6006

Rashedul Hasan is a lecturer in accounting at Coventry University, UK. He completed his PhD in 2018, which focuses on corporate governance issues for a non-profit organisation. He is also a CIMA member and holds the designation of ACMA, CGMA. His research interests include corporate governance dynamics, performance and risk management aspects of financial institutions, sustainability and accountability issues. Rashedul is serving as the Associate Editor of the *Journal of Economic Cooperation and Development*. He has published his research in ABS, ABDC and SCOPUS indexed accounting and finance journals.

Azizul Hassan (PhD) is a member of the Tourism Consultants Network of the UK Tourism Society. His research interest areas include technology-supported marketing for tourism and hospitality, immersive technology application in the tourism and hospitality industry, technology-influenced marketing suggestions for sustainable tourism and the hospitality industry in developing countries. Hassan has authored over 100 articles and book chapters in leading tourism outlets. He is also part of the editorial team of 25 book projects from Routledge, Springer, CABI and Emerald Group Publishing. Hassan is a regular reviewer for several journals. ORCID: 0000-0002-0636-963X

Ozan Kaya (PhD) is an Associate Professor at the Vocational School of Ortaca at Mugla Sitki Kocman University, Turkey. He is head of the Hotel, Catering and Services Department. Ozan's areas of expertise are sustainable tourism, gastronomy and alternative tourism types. ORCID: 0000-0003-1461-0320

Małgorzata Kmak received her Master's degree in 2007 from the Pedagogical Academy in Cracow, majoring in political science with a specialisation in international politics. In 2011 she undertook her PhD studies in the Faculty of Humanities at the Pedagogical University in Cracow, majoring in political science. In 2012 she defended her doctoral thesis at the Institute of Political Science, UP. Małgorzata's academic interests include local government, public administration in Poland and Europe, regional policy, sustainable development, energy security, public management, administration, promotion and territorial marketing.

Agnieszka Legut is an Assistant Professor at the Pedagogical University of Cracow. She holds a PhD in sociology. She is also a linguist and a graduate in European studies from Jagiellonian University. Her postgraduate studies were in foreign service and global development. Agnieszka has undertaken courses run by the UNDP and the Council of Europe and has recently completed a course on EU Asylum and Immigration Law at the ULB, Brussels. Her research interests involve immigration and integration policies. In her work, Agnieszka has focused on the Italian, Polish and European context. She has given guest lectures at the University of Milan, Siena and Messina. ORCID: 0000-0002-4605-5328

Hoang Van Long is currently a research fellow at the Da Nang Socio-Economic Development Research Institute. His doctorate is in agricultural and natural resources economics from Kyushu University, Japan in

2012. Long has published five articles in international journals, one international monograph and 20 proceedings from domestic and international seminars. He is currently the inviting lecturer at the Da Nang University of Economics and Tay Nguyen University. Long was honoured to be invited by two universities in France (University of Nantes and University of Angers) as an invited professor to develop project proposals in the field of economics and environmental valuation.

Mohammad Dulal Miah is an Associate Professor and Head of the Department of Economics and Finance at the University of Nizwa, Oman. He obtained his MBA in finance and his PhD in development economics from Ritsumeikan Asia Pacific University, Japan. Mohammad has co-authored three academic books and contributed more than 30 research papers to peer-reviewed journals. He has attended numerous international conferences as invited speaker and facilitator in various countries. Mohammad's research interests include institutional economics, corporate governance, Islamic finance and banking, comparative financial systems and environmental finance.

Chinyere C.P. Nnorom is a Professor of Demography in the Department of Sociology at Alex Ekwueme Federal University Ndufu Alike (AE-FUNAI), Ebonyi State, Nigeria. Her research interests cover social problems, children, youth, women studies and poverty-related issues. The oppression faced by women in Nigerian societies led to her research on 'Abused: Violence Against Women in Southwest Nigeria'. Chinyere's chapter in this volume, co-authored with Kennedy Eborka, continues to reiterate the need to revisit the empathetic conditions of women instigated by culture.

Chinyere Celine Nwokeke is currently a PhD student in the Department of Sociology at the University of Lagos, Nigeria. Her area of interest is medical sociology with a focus on the impact of sociocultural factors on human health and illnesses. She is a member of the Academic Staff Union of Research Institutions (ASURI), Nigeria, and has a number of publications in both local and international journals. Chinyere is the author of *Dietary Revolution and Herbal Management: A Road Map to the Prevention and Management of Chronic Diseases; Diabetes, High Blood Pressure, Stomach Ulcer, Cancer and Asthma.* ORCID: 0000-0003-0225-8833

Ancuţa Elena Prisacaru is a lecturer in the Faculty of Food Engineering at Ştefan cel Mare University of Suceava, Romania. Ancuţa's areas of expertise are raw food materials quality, food contaminants and food industry technologies, as well consumer behaviours. Currently Ancuţa is working on a concept of innovative foods obtained from industry by-products. She is also an enthusiastic sketcher, mountaineer and bike rider. ORCID: 0000-0001-7400-4002

Renata Rettinger is an Assistant Professor at the Pedagogical University of Cracow, Institute of Geography, Department of Tourism and Regional Studies. She received her MSc and PhD degrees in earth science, in the field of geography, from the Pedagogical University of Cracow. She is a board member of ASAP-Poland. Renata's academic interests focus on issues related to determinants and consequences – as well as the process itself – of tourism development in selected regions of the world, the Caribbean in particular. Her research, in reference to sustainable growth and pro-poor tourism, concentrates mainly on enclave tourism as a prevalent form of tourism development. ORCID: 0000-0003-4393-4340

Urszula Śliwka holds an MSc in human nutrition and is also a registered nutritionist with the Association for Nutrition, UK. Since 2016 she has also been a PhD candidate at the Faculty of Food Science in Poland and a Research Fellow in Women and Children's Services at Jersey General Hospital. Currently, Urszula is the Health Promotion Lead in the Preventive Programmes Team for the Government of Jersey. Her primary interest in this current role is to engage and empower individuals and communities in particular vulnerable subgroups of the population and to make changes that reduce the risk of preventable diseases through community health development projects. ORCID: 0000-0001-5316-7742

Van Canh Truong is a lecturer in the Faculty of Geography at the University of Danang – University of Science and Education, Vietnam. He received his doctorate in socioeconomic geography and spatial management from the Faculty of Geography and Regional Studies at the University of Warsaw, Poland. Truong's areas of expertise are sustainable development and socioeconomic geography issues. He is currently working on the measurement of sustainable development by using indicators, as well as the measurement of city prosperity and quality of life. ORCID: 0000-0001-6024-6692

Barbara Woś is an Assistant Professor in the Department of Tourism and Recreation at Opole University of Technology. Her areas of interest are agrotourim, ecotourism and biodiversity conservation, and she is currently working on the concept of sustainable rural development. Barbara is also a wildlife lover and a promoter of beekeeping and phytotherapy.

Mitsuyasu Yabe is a Professor in the Department of Agricultural and Resource Economics at Kyushu University in Japan. He has 30 years' experience as a researcher, teacher and consultant in the fields of environmental economics, natural resource economics and agricultural economics. He is an expert on environmental valuation and has worked on landscape and biodiversity valuation and on the use of economic analysis in the design of agri-environmental policy. Mitsuyasu has also been involved in a variety of research activities on environmentally friendly agriculture, the effective use of biomass for renewable energy policy and food safety. He has a PhD from the Department of Agricultural Economics, Kyoto University, Japan.

Foreword

Human beings meet their needs through labor – their own and that of others: the labor of mothers, peasants, builders, factory and transport workers, doctors and nurses, educators and researchers, artists and entertainers, engineers and planners. Such productive labor provides goods and services that, directly or indirectly, fulfill human needs.

Then there is predatory labor, labor that does not provide goods or services for the fulfillment of human needs but rather appropriates the fruits of others' productive labor. This includes much of the labor of burglars and warriors, slaveholders and arms dealers, priests and nobles, speculators, investors and wealth managers, lawyers and accountants, politicians and lobbyists. To be sure, such labor can be quite valuable, as when one group of warriors or lawyers fiercely protects your body or rights from another such group. Still, this is hardly a reason for wanting such predatory labor around at all.

As humankind becomes ever more productive, an ever larger share of the fruits of productive labor can sustainably be appropriated by predatory labor. Insofar as this can happen, it usually does happen, because the experts in predation tend to do better in the competition over the social product than its producers. This is so for two systematic reasons. Firstly, they are much better trained and prepared for such competition because redistribution is the center of their professional expertise; for tax lawyers, speculators or lobbyists, it is the essence of their job to bend the distribution of wealth or political influence in their favor or in favor of their paying clients. Peasants, builders, teachers, physicians and engineers, by contrast, are not trained for, or professionally occupied with, such redistribution but rather with producing food, care, learning, buildings, tools and machines. This first reason makes it likely that predatory labor wins superior rewards within most social systems.

Such superior rewards, in turn, bring the second reason into play: winning superior rewards, predatory labor then also attracts a disproportional share of the most creative and energetic people. The brightest young minds then dream of being a priest rather than peasant, banker rather than engineer, lawyer rather than doctor, investor rather than inventor, politician rather than artist. To be sure, some bright young minds will still want to be teachers, artists or scientists even if they could earn much more on Wall Street or K Street.[1] But most will 'sell out' and thereby contribute their talents to further increase the capacity of predatory labor to capture large and increasing shares of the social product. This trend is reinforced by the fact that those who engage in predatory labor have a shared vested interest in using their strong social position for promoting a societal ethos that makes income and wealth the foremost bases of social recognition. Because their predatory labor can hardly be presented as admirable, they favor a culture where people are respected for their income, wealth and spending.

As societies become richer, they become increasingly dominated by those who do predatory labor and increasingly permeated by a culture that assigns social worth on the basis of money.

How dangerous and destructive these two trends are becoming apparent when we reflect on how deeply the values, understandings and emotions of persons are shaped by their professional roles and interests, and on how dramatically the interests associated with many predatory-labor roles diverge from any plausible conception of the common good. Eager to increase their profits and influence, arms dealers are longing for tensions and arms races, for hostilities and wars, as conditions in which their business can thrive – and they make efforts to promote such conditions. To be sure, they do not want an all-out nuclear war that would destroy life on this planet. But they do work towards a world on the brink of war, a world of high distrust in which trillions each year are spent on the military. Such a world comes with a lot of low-level violence and continuous small risks of serious destruction. But, arms dealers perceive these costs and risks as acceptable and present them as necessary for national security.[2]

The professional interest of politicians also gravitates toward hostility and tension. When ordinary citizens perceive their country to be threatened by nasty rivals, they feel insecure and tend to follow and support their political leaders. This tendency to 'rally around the flag' especially benefits the executive branch of government which gains stature against the other branches whenever there is a heightened danger of violence. Even if the top officials of the executive and the armed forces would not benefit from a bloody shooting war, they nonetheless understand that their own political power is enhanced by a vivid awareness that such a war is a real possibility. Damaged political leaders are often sorely tempted to lead their country to the brink of war to save or enhance their position, something that can be done openly but also by stealth or misinformation (creating the false appearance that our country has been wronged or threatened or attacked by others). And the aggressive postures associated with being on the brink of war then make actual war and violence substantially more likely.[3]

Speculators make money on big swings in markets, so they have a professional interest in instability. This interest is initially passive:

speculators try to sniff out especially volatile markets and then build positions in such markets in the direction in which they expect the next big move to occur. Once such a position is built, the interest becomes more active, as speculators are then inclined to look for ways to *make* their chosen market more volatile and to trigger a major market move in their favor. This can be done through rumors, for example, or through clandestine or tacit collaborations designed to move markets. Through such machinations, full-time professional speculators and traders can most easily fleece amateur market participants, such as farmers who sell futures contracts to reduce their market risk or retail investors who own stocks to pay for their future retirement. Big market moves involve massive redistributions from the amateurs to the professionals, as is amply documented in records of trading desk profits. Some such market swings seem to have no comprehensible external reasons – just some professionals finding it in their interest to rock the boat, often in teamwork. In other cases, such external reasons exist, such as the COVID-19 pandemic, for instance, but even there the professionals are seeking to exacerbate profitable volatility, thereby destabilizing markets and often causing severe losses to less savvy market participants. Such unsophisticated participants can include whole states, as is exemplified in the currency markets where some US$7 trillion are traded each day (this volume is 28 times greater than the gross world product, the value of all the goods and services produced worldwide, which currently amounts to about US$90 trillion per year!). Among the countries whose currencies have been devastated by speculative attacks in recent decades, causing vast disruption and impoverishment to their populations, are India (1991), Mexico (1994), Thailand (1997, with Indonesia, South Korea, Laos, Malaysia and the Philippines also dragged in), Russia (1998 and 2014), Ecuador (1998–1999), Argentina (1999–2002), Brazil (1999), Turkey (2001 and 2018) and Uruguay (2002).[4]

As a final example, tax lawyers and accountants have a professional interest that the tax environment be highly complex and ever changing. Pursuing this shared interest, they have been promoting a world with a huge diversity of complex and frequently revised national tax regimes that afford rich individuals and corporations endless opportunities for tax evasion and abuse. To exploit these opportunities, a huge global industry has sprung up of tax havens, secrecy jurisdictions, letterbox companies, fake trusts and anonymous accounts, along with an army of shady lawyers, accountants, lobbyists and financial advisors. This netherworld of global finance is estimated to manage over US$30 trillion in wealth, equivalent to about 9% of all private financial wealth or one-third of the annual gross world product.[5] As recently revealed – by LuxLeaks (2014), Panama Papers (2015), Bahamas-List (2016), Paradise Papers (2017), etc. – this sophisticated infrastructure facilitates not merely massive tax abuse, but also crimes of many other kinds such as illegal trade in persons, drugs and weapons, embezzlement, international terrorism, hacking of democracy, and the money laundering associated with all such activities. In these ways, the global haven industry massively aggravates national and global inequalities and greatly impedes the development of poor countries by enabling multinational corporations, autocrats, corrupt officials, millionaires and criminals to drain them of capital and tax revenues. The NGO Global Financial Integrity estimates that such illicit financial outflows from developing countries amount to US$1.1 trillion *per annum* (GFI, 2020). It also serves to store an immense amount of hidden wealth that can be deployed for corrupt purposes, in particular for corrupting the procedures and officers of states and corporations.

Summing up, I have sought to support two points:

- While the role-specific interests of productive labor tend to be reasonably well aligned with the long-term collective interest of humankind in stable peace, prosperity and ecological balance, the role-specific interests of predatory labor often dramatically diverge from the latter and are also much more short term.
- The governance institutions that determine social organization and infrastructure on all levels – global, regional, national and subnational – are heavily dominated by people in predatory-labor roles, who represent a minority of human beings but a large and increasing preponderance of political power.

These two points help explain and predict increasing economic, social and political inequality with the associated persistence of severe poverty, as well as increasing instability with the associated grave dangers of highly destructive wars and ecological disaster.

One response to this situation is the struggle for democratization with the goal of better aligning governance with the long-term interest of humankind. Progress in this dimension might be achieved through mass mobilization from below or through charismatic and far-sighted political leadership from above. Both of these possibilities look remote, and increasingly so, but also progressively more urgent.

The present book explores two less ambitious but also more practicable strategies. One of these, represented by the chapters in Part 1, aims for small but consequential reforms in governance or infrastructure, seeking at least marginally to improve their alignment with the common good. The other, represented by the chapters in Part 2, aims to limit the reach of predatory labor: to protect local communities from predatory exploitation and disruption.

In times when the kind of deep transformative change envisioned by the first strategy seems unrealistic or even unreal, it is important to explore the other two strategies which can articulate a constructive rejection of the status quo, mobilize and focus political engagement, and also keep alive hope as well as admirable examples of sustainable, democratic and solidary human communities. In this spirit, I commend this book's chapters to your attention.

Thomas Pogge
Yale University

Notes

(1) Wall Street is the symbolic center of the financial district of New York City. K Street in Washington, DC, houses many lobbying firms that seek to influence politicians and so-called civil servants.

(2) Some introductory information about arms industry lobbying in the United States and Europe can be found, respectively, at https://www.huffpost.com/entry/arms-industry-ramps-up-lo_b_7484338 and https://corporateeurope.org/en/power-lobbies/2017/12/arms-industry-lobbying-and-militarisation-eu (accessed 25 May 2020).

(3) The seminal essay on this subject is Mueller (1970). For a later, expanded and more empirical treatment, see Kuijpers (2019).

(4) For a general introduction, see https://www.encyclopedia.com/social-sciences-and-law/economics-business-and-labor/money-banking-and-investment/speculation (accessed 25 May 2020). A more detailed analysis is provided by Di Maggio (2016). On the important special case of food markets, see de Schutter (2010), and also Lagi *et al.* (2011).

(5) For a brief introduction, see https://financialtransparency.org/wp-content/uploads/2016/12/FTC_infographic_20170202_Eng.pdf (accessed 25 May 2020). A credible estimate of $24–36 trillion is provided by Henry (2016: 48, 79). For reasons Henry (2016: 79) gives, the estimate by Zucman seems too low/restrictive (see Zucman, 2015). Aggregate global wealth in mid-2019 is estimated at $351.5 trillion (see Shorrocks *et al.*, 2019: 5).

References

de Schutter, O. (2010) Food commodities speculation and food price crises: Regulation to reduce the risk of price volatility. Briefing Note No. 02. UN Special Rapporteur on the Right to Food. See www.srfood.org/images/stories/pdf/otherdocuments/20102309_briefing_note_02_en_ok.pdf (accessed 25 May 2020).

Di Maggio, M. (2016) Market turmoil and destabilizing speculation. Columbia Business School Research Paper No. 13–80 (1 August. See https://ssrn.com/abstract=2353133 (accessed 25 May 2020).

GFI (2020) Trade-Related Illicit Financial Flows in 135 Developing Countries: 2008–2017. GFI, Washington, DC: Global Financial Integrity. See https://gfintegrity.org/report/trade-related-illicit-financial-flows-in-135-developing-countries-2008-2017/ (accessed 25 May 2020).

Henry, J. (2016) Let's tax anonymous wealth! A modest proposal to reduce inequality, attack organized crime, aid developing countries, and raise badly needed revenue from the world's wealthiest tax dodgers, kleptocrats, and felons. In T. Pogge and K. Mehta (eds) *Global Tax Fairness* (pp. 31–95). Oxford: Oxford University Press.

Kuijpers, D. (2019) Rally around all the flags: The effect of military casualties on incumbent popularity in ten countries 1990–2014. *Foreign Policy Analysis* 15 (3), 392–412.

Lagi, M., Bar-Yam, Y., Bertrand, K. and Bar-Yam, Y. (2011) The food crises: A quantitative model of food prices including speculators and ethanol conversion (22 September). See https://papers.ssrn.com/sol3/papers.cfm?abstract_id=1932247 (accessed 25 May 2020).

Mueller, J. (1970) Presidential popularity from Truman to Johnson. *American Political Science Review* 64 (1), 18–34

Shorrocks, A., Davies, J. and Lluberas, R. (2019) *Global Wealth Report 2019*. Zurich: Credit Suisse Research Institute. See https://www.credit-suisse.com/about-us/en/reports-research/global-wealth-report.html (accessed 25 May 2020).

Zucman, G. (2015) *The Hidden Wealth of Nations*. Chicago, IL: University of Chicago Press.

Preface: An Unequal World

Michal Apollo and Pahlaj Moolio

Understanding Poverty and Development

This book seeks to provide interdisciplinary (nay, rather multidisciplinary) perspectives on poverty and development. The Cambridge Dictionary defines the word 'poverty' in simple terms as the condition of a person being poor. Classically, two major traditions for defining poverty are notable: relative and absolute. Relative poverty is defined as the state of having inadequate resources to maintain an acceptable way of living (Townsend, 1979); absolute poverty is a condition characterised by severe deprivation of basic human needs. The multidimensional definition of poverty was adopted in Copenhagen in 1995. It denotes a state in which poor people experience various conditions in their daily lives, such as poor health, lack of education, inadequate housing, limited access to clean water and healthy food, disempowerment, poor quality of work, the threat of violence, and living in areas that are environmentally hazardous, among others. Thus, poverty can exist in social, economic and political terms (see Pogge, 2010; Sachs, 2015).

It is generally agreed that development deals with the alleviation (or eradication) of poverty. Poverty and development have been interconnected since time immemorial (see, for example, Landes, 1998). It could also be argued that development is associated with improvements – positive changes in poverty levels, income equality and employment (Seers, 1969) or economic growth (Sachs, 2005), for instance. Furthermore, development can also encompass quality of life, which includes objective well-being (e.g. work, leisure, education) and subjective well-being (e.g. values, expectations). Additionally, development can be viewed as a philosophy in which a society looks forward to a desirable state, a process through which that state is achieved, an outcome of that process or the plan guiding the society to attain its desirable state (Sharpley *et al.*, 1992). More broadly, Arbab (1984) argues that development should be a 'people-centered' model for a scientific practice that supports tolerance, sustainability, peace and justice for all.

Contributing to this debate is Amartya Sen's *Development As Freedom*, which moves beyond the dominant developmental paradigm by asserting the need for a new means of interpreting the relationships between a set of linked freedoms: political freedoms and transparency in relations between people, freedom of opportunity and economic protection from abject poverty. Sen (1999: 8) argues that poverty is characterised by the lack of at least one of these freedoms. In essence, Sen takes 'a view of development as an integrated process of expansion of substantive freedoms that connect with one another'. According to Sen, these freedoms are access to healthcare, education, political dissent, economic markets and equality, and each freedom encourages the development of another. Sen depicts the process of the 'expansion of substantive freedoms' as 'both an ends and a means of development', and therefore promotes a dialectical understanding of development, whereby political, economic and social freedoms 'link with each other and with the ends of enhancement of human freedom in general' (1999: 10).

Thus, it is widely agreed that both poverty and development are multidimensional and complex. Additionally, both poverty and development are very broad topics about which much has been said and written (see, for example, Hall & Patrinos, 2012; Holden, 2013; Payne & Phillips, 2010; Pogge, 2008; Sachs, 2015). These topics can therefore be considered so extensive that limiting them to a specific definition would be a big mistake (see, for example, Bourguignon & Chakravarty, 2019).

Similarly, development is often assumed to be socioeconomically, politically or scientifically and technologically biased. Arguably, the concept of development is a complex one. Not only is it difficult to describe, but it is also hard to measure (Oyeshola, 2007). Thus, what does development mean?

Everyone, or almost everyone, is in favour of development. Often, people use this word as a slogan for what they simply consider to be good (Payne & Phillips, 2010). Arndt (1989) notes that if you ask educated citizens from both developed and developing countries what they understand to be the desirable purpose of the development process, you will receive astonishingly different answers. Development, therefore, must be adapted to the type and character of poverty. Most importantly, this development has to be sustainable. Thus, sustainable development is the central challenge of our times (see, for example, Sachs, 2015). Unfortunately, following the rules of sustainable development to fight against poverty is not easy. As Pogge mentions, the system we live under is designed by the rich for the rich (Duarte, 2016), and because of that, the fight is full of traps. However, in order to fight and conquer poverty, we need to focus on current problems and foresee future ones.

Structure and Intention of this Book

This book aims to bring together interdisciplinary (indeed, multidisciplinary) perspectives to broaden our understanding of poverty. We invited authors to submit both empirical and conceptual chapters on a range of topics related to the impact of poverty on sustainable development, including but not limited to the following:

- natural or social changes in the environment caused directly or indirectly by people, which pose an existential risk to humans, especially to those living in countries of the Global South;
- the natural, social and economic aspects of human behaviour and environmental sustainability;
- the impact of global warming on various aspects of human well-being, including immigration and integration policies, especially as seen through local and regional lenses;
- impact assessments of the environmental consequences (both positive and negative) of industry, agriculture and services (e.g. tourism) in terms of natural, social or economic aspects;
- new concepts in tourism development that consider tourism to be one of the key contributors to the prosperity and well-being of all stakeholders, especially in the countries of the Global South;
- overconsumption in the context of human behaviour and its impact on sustainable development.

The book is a collection of 15 chapters by international scholars who represent a wide range of academic disciplines. It is thus not only interdisciplinary but can also provide the reader with a broad perspective on poverty and development. The book is divided into two parts. The first part presents poverty and development issues from a global perspective while the second part offers local perspectives.

The first chapter deals with the management of ecosystem services. Celina Aznarez argues that unequal access to ecosystem services constitutes one of the main environmental problems – a source of ecological–distributive conflict in a society organised around scarce resources. The second chapter, written by Agnieszka Legut, explores climate change induced migration based on a review of the literature and an EU case study. Her analysis shows that there has been no comprehensive policy development at the EU level regarding the phenomenon in question, but rather only references scattered across a number of policy fields, and the widely contested 'climate refugee' term is avoided. Chapter 3 covers another quality that it is essential to consider in the study of modern civilisation: uncontrolled urbanisation. Małgorzata Kmak proves that the slogan of sustainable development, so popular in planning documents, is not implemented in practice at all. In Chapter 4, Urszula Śliwka explains that nutritional deficiencies affect not only the Global South but also the Global North. She argues that overconsumption of calories can occur alongside poor nutrient intake and malnutrition due to a lack of essential micronutrients. The next three chapters concern tourism, which is universally known to be a significant catalyst for economic growth and job creation. However, as Agnieszka Gawlik explains in Chapter 5, contemporary tourism, which is the source of various benefits, causes different kinds of costs to the economy, society and the environment. She proposes an evaluation criterion that should be used in the formulation, realisation and identification of the impact of sustainable tourism policy. Chapter 6 explores the concept of slow tourism. Ozan Kaya shows that slow tourism has great potential for sustainable development and poverty reduction through regular income distribution, low-carbon travel and accommodation, and practices that ensure/support the use of local production and agriculture. Chapter 7, co-authored by Agnieszka Gawlik and Barbara Woś, describes agritourism and ecotourism as innovative directions for sustainable rural development. The authors show that agri- and ecotourism generate significant contributions to the process of sustainable rural development with specific effects on the environment, agricultural heritage or economic growth. Chapter 8 again delves into agriculture; however, this time the perspective is economic. The work of Rashedul Hasan, Mohammad Dulal Miah and Muhammad Ashfaq provides valuable findings for emerging economies and specifies policy recommendations by endorsing the role of central banks to implement an agriculture-friendly credit policy to maintain sustainable economic development.

The second part of the book contains seven chapters that focus on specific case studies; all of the research was preceded by insightful background work. Chapter 9, by Van Canh Truong, focuses on measuring sustainable development. Using Vietnam as a case study, Truong stresses that a proper evaluation of sustainable development is needed to meet the needs of the present without compromising the ability of future generations to meet their own needs. In Chapter 10, Cristina Ghinea and Ancuţa Elena Prisacaru look at sustainable consumption based on a survey conducted in Romania. The authors indicate that there was a high level of knowledge about sustainable consumption among respondents, but a moderate attitude and practice level. Thus, they recommend moving from theory to practice. Chapter 11 focuses on environmental sustainability concerns. By investigating the environmental consequences of the recent oil spill in the world's largest mangrove forest – the Sundarbans in Bangladesh – Azizul Hassan explains and analyses business models that address environmental sustainability. In Chapter 12, Kennedy Eborka, Chinyere C. Nwokeke and Mosope Akinbola deal with the negative consequences of urbanisation. The researchers from Lagos focus on access to affordable housing and sustainable development among low-income earners in Nigeria. Based on their recommendations, governments should shift from the usual elitist approach to the provision of housing and place greater emphasis on housing units for low- and medium-income earners. Chapter 13 returns us to Nigeria, and Chinyere C.P. Nnorom and Kennedy Eborka discuss the structure of patriarchy and its applications for the exploitation, deprivation and subjugation of females in family and community life. The authors recommend compulsory education for girls up to secondary school level, the strengthening of existing legal codes to protect females from all forms of abuse and exploitation, and the introduction of social security for non-working widows and elderly and unemployed women. In Chapter 14, Renata Rettinger explores the present and potential environmental, societal and economic conflicts brewing in tourist regions in the context of sustainable development. Using Cuba as a case study, she claims that, despite the negative issues, tourist enclaves will continue to be built on islands or in spatially isolated geographic areas. Chapter 15, written by Hoang Van Long and Mitsuyasu Yabe, explores environmental factors affecting agricultural production. The authors stress that plant disease, soil fertility, irrigation management and water pollution have negatively affected profit in rice production in Vietnam. In short, farmers should reduce their household size and the number of farm plots, apply a row-seeding technique and improve environmental factors to obtain high-profit efficiency.

Despite the different contexts and discourses of these contributors, each case highlights poverty and development. However, it must be mentioned that it is impossible for even a comprehensive publication to discuss every single challenge that the current world faces. It is also not possible to indicate all future prospects. Nevertheless, the issues presented here are worthy of attention, even if they comprise only a fraction of the whole picture. The works included here are an attempt to cast some light on some dimensions of these topics.

References

Arbab, F. (1984) *Rural University: Learning about Education and Development*. Ottawa: International Development Research Centre.

Arndt, H.W. (1989) *Economic Development: The History of an Idea*. Chicago, IL: University of Chicago Press.

Bourguignon, F. and Chakravarty, S.R. (2019) The measurement of multidimensional poverty. In S.R. Chakravarty (ed.) *Poverty, Social Exclusion and Stochastic Dominance* (pp. 83–107). Singapore: Springer.

Duarte, M. (2016) Stand against poverty: An interview with Thomas Pogge. *Etikk I Praksis* 2, 91–108.

Hall, G.H. and Patrinos, H.A. (eds) (2012) *Indigenous Peoples, Poverty, and Development*. Cambridge: Cambridge University Press.

Holden, A. (2013) *Tourism, Poverty and Development*. New York: Routledge.

Landes, D. (1998) *The Wealth and Poverty of Nations: Why Some Are So Rich and Some So Poor*. New York: W.W. Norton.

Oyeshola, D. (2007) Development and poverty: A symbiotic relationship and its implication in Africa. *African Journal of Traditional, Complementary and Alternative Medicines* 4 (4), 553–558.

Payne, A. and Phillips, N. (2010) *Development*. Cambridge: Polity Press.

Pogge, T. (2008) *World Poverty and Human Rights*. Cambridge: Polity Press.

Pogge, T. (2010) *Politics as Usual: What Lies Behind the Pro-Poor Rhetoric*. Cambridge and Malden, MA: Polity Press.

Pogge, T. (2013) Poverty and violence. *Law, Ethics and Philosophy* 1, 87–111.

Sachs, J.D. (2005) *The End of Poverty: Economic Possibilities for Our Time*. New York: Penguin.

Sachs, J.D. (2015) *The Age of Sustainable Development*. New York: Columbia University Press.

Seers, D. (1969) The meaning of development. *International Development Review* 11 (4), 3–4.

Sen, A. (1999) *Development As Freedom*. New York: Oxford University Press.

Sharpley, A.N., Smith, S.J., Jones, O.R., Berg, W.A. and Coleman, G.A. (1992) The transport of bioavailable phosphorous in agricultural runoff. *Journal of Environmental Quality* 21, 30–35.

Townsend, P. (1979) *Poverty in the United Kingdom*. London: Allen Lane and Penguin.

Part 1

Poverty and Development: A Global Perspective

1 Social Distribution of and Access to Ecosystem Services: Approaches for the Alleviation of Socio-environmental Inequalities

Celina Aznarez

Ecosystem services (ES) and environmental degradation are unequally distributed in societies. Those who manage and decide on ecosystems politically mediate the processes that determine ES production and distribution. Therefore, conflicts arise between the beneficiaries of ecosystems and those affected by their degradation. Different conceptual frameworks are emerging as attempts to analyse the contributions of ES to poverty alleviation. In this chapter I analyse different theoretical contributions about the production and social distribution of ES with regard to the socio-environmental conflicts arising from their distribution. The lack of consensus on decision making at different scales (e.g. geographical, temporal, social), as well as criteria between the evaluation of ES and their management, are serious limitations to the application of the ES framework to decision making. Revised contributions highlight that there is a growing need to incorporate multi-scale analyses of how ES are generated and the socio-ecological processes that model their production and social distribution. Efforts should focus on identifying the mechanisms through which decision making can be an aggravating circumstance in situations of vulnerability or poverty. Integrated into practice, the discussion of value chains generated around ES would recognise the value associated with ecosystems as multidimensional. In this way, dimensions that make the distribution of ES a mechanism for poverty alleviation and human well-being would be incorporated into decision making. Given that the many contributions from different thematic areas are currently disjointed, the need to articulate visions from different disciplinary fields in an adapted way is essential for management.

1 Introduction

Human societies depend on ecosystems to fulfil the different requirements for human well-being: access to security, basic materials for a decent life, health, good social relations and freedom of choice/action (MEA, 2005; Narayan *et al.*, 1999). The complex interactions between human societies and the natural environment define the so-called socio-ecological systems (SES) (Berkes *et al.*, 2008; Janssen & Ostrom, 2006; Walker *et al.*, 2004) that regulate the flow of goods and ecosystem services (MEA, 2005; Ghermandi *et al.*, 2013). The Millennium Ecosystem Assessment (MEA) introduced a new framework for analysing SES, considering the ecosystems through the lens of service providers and their relationships to human well-being (Carpenter *et al.*, 2009). In this context, ecosystem services (ES) are defined as all of the flows of an ecosystem that, produced naturally, result in relatively immediate benefits for humans (Brown *et al.*, 2006; Daily, 1997; MEA, 2005). These can materialise directly as supply services (provision of food and natural resources, such as wood and water) or cultural services (recreational, spiritual and religious benefits); or indirectly through regulation services (of floods, droughts, soil degradation and diseases) and support services (soil formation and nutrient cycling) (MEA, 2003). However, obtaining these benefits frequently coincides with non-linear changes or increasing degradation in ecosystems, affecting their ability to provide ES (Butler *et al.*, 2003; Carpenter *et al.*, 2006; MEA, 2005).

Social dependency on ecosystems accentuates the relevance of promoting norms and actions aimed at sustaining the capacity of ecosystems to provide certain ES (Ernstson, 2008). These ways of managing ecosystems are the means by which societies also define the social distribution of ES and, therefore, who can benefit from them (Ernstson, 2008). Different social processes linked to ecosystem management govern access and control to ES, affecting how the users of said services compete for or are excluded from access to exhaustible services or those that may become unavailable (Fisher *et al.*, 2014; Ribot & Pelusso, 2003). According to Costanza (2008), in cases where actors linked to the use of an ecosystem can regulate the access to or availability of given ES for other potential users, the concepts of private property and markets linked to this ecosystem arise. This can influence the emergence of conflict dynamics among the users of the ES, who are able to dispute the appropriation of a specific service or condition each other through differential access or depletion (Fisher *et al.*, 2014).

This differential access to ES can manifest from the approach of distributive ecological conflicts (Butler & Oluoch-Kosura, 2006; Ernstson, 2013). It then becomes necessary to develop interdisciplinary approaches that deepen understanding of distributive conflicts and the mechanisms that promote these distributive processes giving origin to diverse inequalities. The normative concept of environmental justice is a framework of great interest for the generation of contributions linked to the socio-spatial distribution of goods and ES, including the notion of well-being, as well as the affectations derived from anthropic activity (Ernstson, 2013; Low & Gleeson, 1998). Environmental management based on ES requires knowledge about how the assessment of ES relates to the access and dependence of the stakeholders linked to their use (Rincón-Ruíz *et al.*, 2014). In this sense, the theoretical framework of ES in conjunction with the novel environmental justice framework provides a robust analytical reference from which to generate contributions for the design of equitable environmental management policies regarding well-being and access to ES.

2 Poverty Traps and Vulnerability

ES are related to components of human well-being, and poverty refers to situations where there is a lack of well-being (Daw *et al.*, 2011; Fisher *et al.*, 2013, 2014; MEA, 2003). Environmental vulnerability in poverty is associated with the reduced capacity to face risks, environmental disturbances or changes in the ES provoked by the activities of other social sectors (Boyce, 2007; Butler & Oluoch-Kosura, 2006; Tallis *et al.*, 2011). This implies that the convergence of stressors simultaneously or cumulatively can lead to the inability of people living in poverty to recover by leading already fragile livelihoods into poverty traps (Tschakert, 2016). People living in poverty are economically limited in their ability to acquire market goods that replace natural losses when there are changes in the provision of ES (Boyce, 2007; Butler *et al.*, 2003). Hence, natural capital can help households recover after natural disasters, constituting a safety net for stressful situations (Bharucha & Pretty, 2010; McSweeney, 2005; Paumgarten *et al.*, 2018; Scherr, 2000).

Analysis trends indicate that poverty traps entail complex links between asset poverty, lack of opportunities for access to land, labour and credit markets, as well as deficient access to natural resources and ES (Barbier, 2010; Foladori, 2002). Several authors warn about the connections between macro-economic and political processes imposed by developed countries as the main drivers of change and environmental degradation in the ecosystems of developing countries (Barbier, 2010; Butler & Oluoch-Kosura, 2006; Martinez-Allier, 2002). Among these, a vicious cycle of factors described as poverty traps reinforces the current asymmetries (Butler & Oluoch-Kosura, 2006). In their connection to access to ES, poverty traps are expressed in diseases caused by environmental pollution, food insecurity, poor governance and a lack of education, information, credit and transportation (Bonds *et al.*, 2009; Enfors & Gordon, 2008; Robards *et al.*, 2011; Sánchez-Moreno & Ferris, 2007). Furthermore, other mechanisms preserve the inequalities derived from the status quo, such as misdirecting public discussion in a deliberate and organised way (Apollo, 2020) or through the 'system justification' theory (Apollo, 2020; Jost, 2010). This theory states that people in all existing systems tend to justify 'the way things are' and therefore avoid alternatives to the status quo for ideologically defensive reasons (Apollo, 2020; Jost, 2010). There is a consensus in the literature that attributes consumption patterns of the wealthiest sectors as causes of the relationship between environmental degradation and poverty (Adams, 2008; Boyce, 1994; Foladori, 2002; Lade *et al.*, 2017). According to Boyce (1994), power and wealth inequalities lead to greater environmental degradation for reasons linked to asymmetries in the social weighting of decision making and impacts due to the valuation of costs and benefits of polluting activities as well as the impact of the time preference rate applied to the environment.

The study of systemic vulnerabilities that generate deficiencies in the adaptation capacities of certain groups in times of crisis is crucial. Considering the relevance of ES-based safety nets and incorporating them into planning and decision making to avoid the deprivation of well-being is closely related to poverty prevention.

3 Ecosystem Services in Poverty Alleviation

The conceptual simplification that aggregated ES with well-being and humanity in general terms has led to the belief that the conservation of the different flows of ES results in general welfare, thus alleviating poverty (Daw *et al.*, 2011). However, the benefits that individuals obtain from ecosystems are dependent on a complex network of access and control mechanisms (institutions, capacities, social relations, rights and forms of capital) (Fisher *et al.*, 2014; Ribot & Pelusso, 2003). Therefore, the well-being of the poorest sectors depends on the existence of access mechanisms so that they can benefit from ecosystems (Butler *et al.*, 2003). In this way, increased availability and flow of ES may not generate changes in the well-being of people in poverty, but modifications in access mechanisms, independent of the changes in the ecosystem functions, may do so (Daw *et al.*, 2011).

Defining aggregation as the consideration of the benefits associated with ES for the total of humanity, without explicitly recognising their distribution patterns, Daw *et al.* (2011) identify four key aspects of how ES relate to poverty alleviation, which are overlooked in theoretical–conceptual approaches due to the predominance of the conceptual aggregation of human well-being. Firstly, an aggregated analysis may not consider the winners/losers of well-being in terms of benefits obtained from ES. Secondly, it excludes the identification and study of the mechanisms that determine access to ES and, therefore, their contribution to the well-being of the poor sectors. Thirdly, together with the second point, it overlooks the individual circumstances determining the contribution of ES to well-being. Lastly, it can lead to erroneous conceptualisations of ES and benefits, neglecting the critical mechanisms of poverty alleviation, such as ES-based employment affecting the poverty–ES link approach. Therefore, adaptation and integration of conceptual frameworks linked to ES and well-being are needed to understand the different dimensions that contribute to poverty alleviation. A recent effort towards that goal is Ecosystem Services for Poverty Alleviation (ESPA), an approach that integrates two conceptual frameworks that are complementary to each other. The first of these is socio-ecological systems, which arise from the integration of natural and social sciences as complex systems, associated with concepts of dynamism, non-linearity and uncertainty (Janssen & Ostrom, 2006), understood as inherent characteristics of the systems to be studied and managed (Carpenter *et al.*, 2009). The second is the political economy framework, given the need to incorporate dimensions linked to power, politics (Cote & Nightingale, 2011) and human agency capabilities (Brown & Westaway, 2011). Thereby, the ESPA framework tends to reconsider and develop areas that are still poorly understood by the MEA for poverty alleviation. To understand the links between environmental degradation, ES and poverty alleviation, efforts should focus on responding to the question of *how asymmetries in the weighting of ES decision making affect their production and social distribution.*

4 Social Production and Distribution of Ecosystem Services

Studies, mainly from urban ecology, show the influence of urban development patterns in the modelling of ecological processes generating ES (Alberti, 2005; Pickett *et al.*, 2008). Human presence heavily alters and regulates not only the generation of ES but also their social distribution (Ernstson, 2008, 2013; Lakerveld *et al.*, 2015; Mullin *et al.*, 2018). Social factors contributing to the maintenance of the ES, as well as their distribution, constitute the social production of ES (Ernstson, 2008, 2013). According to Ernstson (2008), the combination of equity and ecological complexity remains poorly understood, issues that could be addressed based on two assumptions. The first is that interlinked socio-ecological processes in interaction with several scales generate ES (Berkes *et al.*, 1998). In urban contexts, these processes connect to the dependence on ecosystems spatially defined by urban development as well as their respective environmental management by society (Andersson *et al.*, 2007; Ernstson, 2013; Mullin *et al.*, 2018). In this way, the social production of ES depends on the ability to develop management practices that sustain ecological flows and networks (Ernstson, 2013). It is also affected by decisions about which ES to prioritise in managed areas, and how these benefit people. Thus, ES generation processes are modelled in all landscapes through patterns derived from socio-ecological structures (Alberti, 2008;

Ernstson, 2008; Marzluff *et al.*, 2008), which helps explain the unequal distribution of ES in different spatiotemporal scales.

The second assumption is that ES tend to generate benefits and value chains at specific scales, such as those in the present and at a local level (e.g. fishing in a lake) and others at a larger scale (e.g. maintaining biodiversity) (Ernstson, 2013). Most benefits obtained from ecosystems are the result of social processes involving capital and work in interaction with ecosystems, and are mediated by institutions such as property rights (Lakerveld *et al.*, 2015; Lele, 2013). There is a key distinction in the analysis of ES production and control between ES that are physical entities (provisions and cash derived from the ES) and a second group that includes the ES of regulation, support and culture (Fisher *et al.*, 2014). The first group is more susceptible to control since these ES often circulate on tangible production chains in the shape of raw materials or tradable products. The second group is more difficult to control since larger spatial scales and complex socio-ecological synergies influence its components (Fisher *et al.*, 2013, 2014).

Frequently, the benefits or values that people obtain from ecosystems become a focus of discussion because of their subjectivity. In such cases, the discussion requires the use of social practice to explain and describe these values and benefits (Ernstson, 2013; Sörlin, 1998). The actors involved in processes of value articulation are frequently artists, authors and scientists. Other actors and social arenas (e.g. media, assemblies and parliamentary exhibitions) can use the resulting productions as tools to argue the value of a phenomenon (Ernstson, 2013; Ernstson & Sörlin, 2009). This is a key aspect, given that cultural differences emphasise the divergence between achieving a combination of ES through environmental management while contemplating the demands from individuals, communities and societies (Robards *et al.*, 2011). Likewise, conflicts arising from the opposition of individual claims over collective objectives lead to incompatible collective results. At this level, the concentration of economic capital is usually associated with a greater political scope and ability to generate mechanisms of self-reinforcement, favouring a greater concentration of wealth, breaking the political process of democracy and ultimately contributing to a climate of social tension (Scheffer *et al.*, 2017; Stiglitz, 2012). Thus, social institutions must allow ES production and distribution to be incorporated into the management dimensions related to the social production of ES (Ernstson, 2008, 2013).

5 Trade-offs and Emergence of Ecological–Distributive Conflicts

Many studies highlight that societal ES management decisions drive the existence of trade-offs between the demand for particular ES affecting the availability of other dependent ES (Chaudhary *et al.*, 2018; Lele, 2013;

Rodríguez *et al.*, 2006). The MEA framework considers trade-offs to be situations where certain stakeholders lose access to ES to the extent that other stakeholders appropriate them or deteriorate them through use. Furthermore, the sustainability of ES is associated with distributive dynamics (Norgaard, 2010). This is linked to existing trade-offs between social groups who access ES and others who are excluded (Chaudhary *et al.*, 2018; Lele, 2013; Robards *et al.*, 2011). There are implicit trade-offs in particular configurations of ES that, in their fulfilment, benefit previously determined social groups (Ferraro & Hanauer, 2014; Robards *et al.*, 2011). Decision making and environmental management are key elements in these dynamics, which often focus on the immediate demands of ES, compromising the availability of those or other ES in the future (Carpenter *et al.*, 2009). Often, prioritising particular ES in terms of their scale and objectives over others can lead to the generation or increase of asymmetries in the access to and use of ES (Ferraro & Hanauer, 2014; Robards *et al.*, 2011).

The conceptualisation of ecological–distributive conflicts or conflicts of environmental justice arises from the postulate of the unequal allocation by individuals regarding the use that the economy makes of the environment (Escobar, 2006; Martínez-Alier, 2003). This concerns interaction processes where different value schemes arise regarding the management of biodiversity, natural resources and ES among social actors mobilised by a shared interest (Martínez-Alier, 1995, 2008; Rincón-Ruíz *et al.*, 2014; Santos, 2010, 2011). These are displayed in the form of economic, cultural and subsistence values expressed at different scales (Martínez-Alier, 2008). It is because of the existence of this diversity of value schemes that the 'ecological–distributive conflict' concept is established, where often the conflict itself escapes the economic logic of awarding a monetary value to the resource in question (Escobar, 2006; Martínez-Alier, 2009). Ecological economics explains that the valuation of nature depends on a series of political and ecological processes that cannot be translated into a market price (Martínez-Alier, 2008). The emergence of these conflicts leads to different stakeholders and language holding different levels of power, culture, knowledge, interest and value.

An unequal wealth distribution and economic development scenarios are often detrimental for environmental quality. Therefore, it is common to see resistance from social groups around the protection of natural heritage (Santos, 2011). Environmental movements can be framed as the 'ecologism of the rich' and the 'ecologism of the poor' (Chaudhary *et al.*, 2018; Martínez-Alier, 2003; Santos, 2010, 2011). The first aims to preserve the landscape and fauna, with a discourse linking the increase in consumption and waste production with biodiversity loss and decreased quality of life (Santos, 2010, 2011). On the other hand, the 'environmentalism of the poor' is a survival strategy and not so much a claim of loss of quality of life (Martínez-Alier, 1995). These conflicts are particularly common in resource-dependent communities that defend the environment as the sustenance for their way of life in the face of a worsening ES distribution without the compensation of greater equality in economic distribution (Martínez-Alier, 1995). This claim has its foundation in the fact that, like ecosystems, human communities have certain requirements for the conditions for their development and survival (Escobar, 1999, 2006). Value schemes for these communities' environments may differ and are highly influential in environmental health. Faced with these ecological–distributive conflicts, when designing better environmental governance, it is crucial to consider *whom this management and environmental planning will benefit.*

6 Environmental Management of the Territory and Ecosystem Services

The lack of agreed principles for decision making at different scales, as well as the lack of integration of criteria between ES valuation and management, have been serious limitations to the efforts to apply the ES framework to decision making (Armsworth *et al.*, 2007; Rincón-Ruíz *et al.*, 2014; Seppelt *et al.*, 2011). Because of these limitations, different methodological proposals that emphasise the link between ES assessment and decision making have been promoted (Chan *et al.*, 2016; Cowling *et al.*, 2008; de Groot *et al.*, 2010; Rincón-Ruíz *et al.*, 2014). The adoption of these proposals implies the development of social systems that allow adaptive management, through the recognition of the value associated with biodiversity and ES as multidimensional, which entails an effort to integrate different language (Chan *et al.*, 2016; Martínez-Alier, 2007). Value language changes over time, transforming itself along with decisions on land use and occupation (Müller *et al.*, 2010). Within this dynamism, it frequently happens that the same territory has attributed different value language and shared interests on the part of different stakeholders, which can compete and generate potential ecological–distributive conflicts (Martínez-Alier, 2005; Tapella, 2012).

In this context, the Integral Assessment of ES and Biodiversity (IAESB) emerges as an adaptive learning process and social empowerment around territories (Rincón-Ruíz *et al.*, 2014). The IAESB incorporates recognition of the ES, with identification and evaluation of the ecosystem functions that support them, also including their benefits and distributive mechanisms through the strategic construction of territorial scenarios (Rincón-Ruíz *et al.*, 2014). This approach allows the integration of an environmental justice perspective to its analysis by including sectors excluded from access to ES, incorporating notions of equity and vulnerability (Prato, 1999). The objective of this practice is to contribute to territorial environmental management through the inclusion of multiple values linked to ES that demonstrate their importance for society. This approach also seeks to analyse the change drivers that affect ES flows, incorporating the construction of possible scenarios for the definition of management instruments of SES. It also contributes to conflict identification, through the disclosure of the different trade-offs and synergies arising from valuation differences (Martín-Lopez & Montes, 2010; Rincón-Ruíz, 2014). The social assessment of ES is then part of the process of ES analysis, with potential to contribute to the understanding of the asymmetric ES distribution among different social sectors (Martínez-Alier, 2003; Tapella, 2012).

There are recent advances in approaches to the environmental management of ES, considering distributive differentiation and environmental justice, based on systemic analysis. To overcome the management challenges outlined previously, several works (Cowling *et al.*, 2008; Rincón-Ruíz *et al.*, 2014) identify five main stages to follow in an integrated assessment of the ES process (Table 1.1). The first two stages are preliminary approaches towards the SES to be managed. They provide analysis of the primary situation and the existing dynamics of the territory governance, as well as the social stakeholders that could participate in the process. The third stage links the ecological values obtained from understanding and quantifying the biophysical processes that provide ES and the sociocultural values that are granted to them. Unlike biophysical valuation, sociocultural valuation is linked to ES demand, as well as to perceived realities and individual and group belief systems (Brondízio *et al.*, 2010; Rincón-Ruíz *et al.*, 2014). For these proposals to contribute to decision making, the identification and analysis of opposing traits or trade-offs that can generate or deepen ecological–distributive conflicts are key (Rincón-Ruíz *et al.*, 2014). In the final stage, the scenario projection strategy is of great importance to integrate the information obtained throughout the IAESB process. The integrated use of tools such as those present in the theoretical framework of IAESB could mean an advancement towards the consideration of the equitable distribution of ES and human well-being in decision making. It would also allow further progress towards the applied integration of the environmental justice framework for ES management through disaggregated analyses (Daw *et al.*, 2011) and the potential to avoid socio-environmental injustices through trade-offs.

7 Environmental Justice and Inclusion of the Ecosystem Services Framework

The social movement of environmental justice originated in the United States at the end of the 1970s. Its objective is to redirect attention from the more traditional approaches linked to environmental conservation, towards the quality of life of the most vulnerable groups of society (Byrne *et al.*, 2002; Mitchell & Norman, 2012; Santos, 2011; Schlosberg, 2007).

Table 1.1 Phases that make up the IAESB methodology

Phase		Description
Preparing the operating model		Preliminary identification of conflicts and scenarios to develop later. Identification of social actors with whom to work during the process. Preliminary definition of research strategies and methodology.
Socio-ecological system characterisation (4 stages)	1. Identification and characterisation of ES	Recognising the provider functions and beneficiary units of ES as well as the language linked to their importance and dependence.
	2. Stakeholder analysis and governance	Knowing the roles of the stakeholders related to the ES as well as the relationship scales. Key knowledge for subsequent analysis of trade-offs.
	3. Socio-ecological trajectory	A historical socio-ecological description that determines relationship patterns between society and nature. Historical description of environmental and territorial management.
	4. Characterisation and definition of the SES	Determination of spatial and functional limits, criteria of cultural identity, monitoring of structure, patterns and temporal processes at the landscape scale. Systematisation of key management factors that structure and compose the SES as well as access and control to ES.
Sociocultural + monetary + ecological assessment phase		Inquiry about perception and relationships regarding ES, as well as power asymmetries in access to them. Parts of the cultural values incorporated as monetary valuation (e.g. values of use + not use). Identification of ES provider functions (ecosystem processes and functional diversity with the capacity to provide ES).
Trade-offs analysis		Differentiated analysis of trade-offs between ES, stakeholders, spatial scales and time scales.
Scenario analysis		Quantitative, qualitative or mixed scenarios. The mixed scenarios complement and mitigate the weaknesses of the other two scenarios separately.

Source: Elaboration from Rincón-Ruíz *et al.* (2014), Cowling *et al.* (2008), Martín-López *et al.* (2012), Martín-López *et al.* (2010), and Vilardy and González (2011).

Originally, the movement arose from mobilisations against what was called 'environmental racism', in which areas inhabited by ethnic groups generally overlapped with environmentally degraded areas, as well as open-air dumps and polluting industries (Bullard, 1990; Chaudhary *et al.*, 2018; Harvey & Braun, 1996; Martínez-Alier, 2012; Santos, 2011). Moreover, we use the term to refer to spontaneous mobilisations and organisations focused on pollution, extractive industries and the consequences of climate change (Martínez-Alier, 2002; Schlosberg, 2013). This approach to environmental problems is becoming a central element in decision making regarding public policies and conservation initiatives (Forsyth & Sikor, 2013).

Recently, environmental justice approaches have begun to consider the equitable distribution of environmental goods and services in connection to well-being and human relations (Ernstson, 2013; Mullin *et al.*, 2018; Wolch *et al.*, 2014). However, due to its complexity and relative novelty, the environmental justice framework has scarcely been incorporated into the ES discourse (Chaudhary *et al.*, 2018). MEA (2005) refers to the subject, mentioning 'freedom of choice and action' as constituents of well-being. This reference does not delve into dimensions such as well-being disaggregation in terms of accessibility to ES, the injustices derived from trade-offs and the environmental burdens from the generation of disservices (understood as the negative relations between the functioning of ecosystems and human well-being) (Chaudhary *et al.*, 2018; Lele, 2013; Sikor, 2013). Despite this, several authors have contributed in this direction. Some have highlighted the need to perform disaggregated analyses regarding ES as a condition for the study of issues related to justice and equity (Daw *et al.*, 2011; Fisher *et al.*, 2013, 2014; Lele, 2013). Others have focused on the differentiated contribution of ES to poverty alleviation and the factors that mediate ecosystems to well-being and the associated distributive inequalities (Lakerveld *et al.*, 2015). In this context, environmental justice has ample potential for the prevention of injustices in the trade-offs derived from environmental management and planning as well as for analysis of how people benefit from ecosystems (Chan & Satterfield, 2013; Lele *et al.*, 2013; Rodríguez *et al.*, 2006; Sikor *et al.*, 2014).

The omission of distributive patterns in the analysis implies that issues related to environmental justice conflicts are ignored and it undermines the efforts to achieve fairer and more equitable societies through ES management (Chaudhary *et al.*, 2018; Mitchell & Norman, 2012). Distribution, participation and recognition are the three fundamental dimensions for environmental justice (Figure 1.1) (Chaudhary *et al.*, 2018; Martin *et al.*, 2014; Schlosberg, 2004, 2013). Within this approach, we understand distribution to be a fair distributive scenario between different social groups, and participation to be a necessary process before fair distributive results and recognition of different cultural identities and values. In this perspective, environmental justice cannot be based merely on distributive

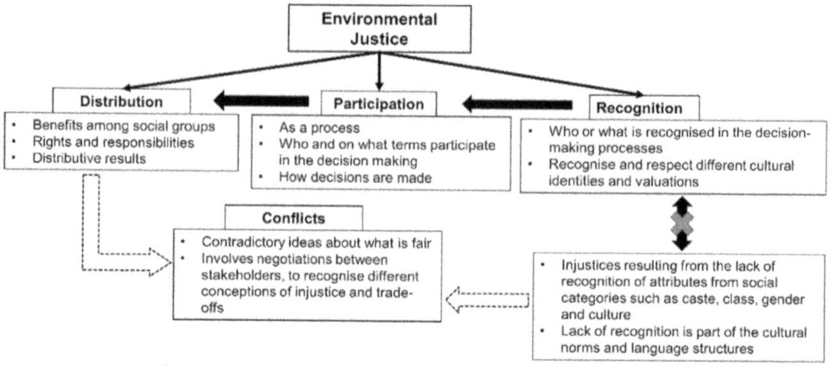

Figure 1.1 Main explanatory dimensions of environmental justice in the analysis of unequal distribution of ecosystem services

Source: Elaboration based on Bohman (2007), Chaudhary *et al.* (2018), Ernstson (2013), Gustavsson *et al.* (2014), Lele (2013), Mitchell and Norman (2012), Schlosberg (2004, 2013) and Sikor (2013).

analyses of ES. It must incorporate the institutional contexts that regulate the participation processes, the cultural attributes that regulate social relations and, finally, inequitable ES distribution (Chaudhary *et al.*, 2018). Furthermore, the dimensions linked to cultural recognition and political participation are fundamental components of the social movement of environmental justice (Schlosberg, 2004, 2013). The aspects that regulate social relations allow us to understand, through this approach, those mechanisms that control the distribution of ES in society as well as the emergence of conflicts around this phenomenon.

8 Scope for Theoretical Integration in Decision Making and Planning

The developed theory around ES, exploring how we distribute benefits obtained from ecosystems, can support environmental management and territorial planning (Jobbágy *et al.*, 2011), thus preventing the multiple asymmetries around access (Fisher *et al.*, 2014; Rincón-Ruíz *et al.*, 2014). In the case of the ESPA conceptual framework, its potential applies to the generation of long-term processes, particularly in research and policy planning. Its application as a tool that contributes to a holistic understanding of socio-ecological and political dynamics is especially useful in cases of resource-dependent populations based on ES (Fisher *et al.*, 2014). It also allows for analysis of the mechanisms that strengthen social differentiation strategies in ES production and distribution. The normative positioning of the ESPA framework, as a promoter for social equity and environmental sustainability, highlights its potential for use in conjunction with the conceptual framework of environmental justice based on ES. This implies the generation of disaggregated ES analysis. Likewise, it has the potential to avoid injustices derived from trade-offs and includes ES distributive dimensions for poverty alleviation and human well-being.

The articulation of valuation methods and approaches, such as those presented in the framework of IAESB, constitute another approach with great potential for research on ES and analysis of socio-ecological dynamics for decision making (Rincón-Ruíz *et al.*, 2014). The political dimension in this approach includes necessary aspects to be taken into account by governmental processes so that the decision making incorporates information about the environmental management of the territory (Jobbágy *et al.*, 2011). It is important to consider the previous relationship dynamics in the SES within the area to be managed as well as the impact of an ecosystemic change on the well-being of the population included (Chan *et al.*, 2016). From the perspective of local communities, incorporating vulnerability dimensions into the analysis is essential to account for adaptive rigidity facing changes in the provision of ES, productive modes and local dynamics. It is in these cases that the performance of the decision makers may be key to not pushing these sectors towards poverty traps and degenerative social cycles. The application of integral assessment methodologies of ES for environmental management and territorial planning would enable the integration of different approaches and areas of knowledge. Doing so would allow prioritisation of different ways of understanding the territory, incorporating the local appreciation in combination with science-based approaches (Chan *et al.*, 2016).

Environmental assessments for land use planning have characteristics of scale and complexity that demonstrate the lack of adaptation of the classical evaluation techniques to the reality to be evaluated. Among them are the Environmental Impact Assessments (EIA), originally designed to quantify the impact of a project or action to be implemented (Valpreda *et al.*, 2003). Currently, these do not incorporate changes in the provision of ES (Jobbágy *et al.*, 2011) nor the access mechanisms. The latter is extremely important in the evaluation of the environmental consequences of the implementation of territorial policies, plans or programmes through the Strategic Environmental Assessments (SEA) (Jobbágy *et al.*, 2011; Partidario & Jesus, 2003). The SEA is a systematic and progressive process, with early incorporation of environmental considerations in decision making, unlike the EIA. In an SEA process, the environmental quality of a policy, plan or programme is evaluated along with the integration of alternative proposals (Partidario & Jesus, 2003). This process also highlights the potential effects that the limitations and opportunities of the medium can exert on the implementation scenarios to be evaluated (Jobbágy *et al.*, 2011). Thus, IAESB and the management scenario evaluation could complement this key tool in decision making. In this way, it would be possible to take into account, prevent and modify some of the access and control mechanisms that result in situations of inequality around ES.

To incorporate these approaches into decision-making tools, a series of elements must be considered. In this process, the biophysical assessment of ES provision (which corresponds to the 'socio-cultural + monetary + ecological assessment' phase of the IAESB) is key. This includes the identification and quantification of the ecosystem functions that support ES and connect to specific benefits (Fisher *et al.*, 2009; Paruelo *et al.*, 2011). The production functions of ES, as well as the beneficiary units, can arise from this analysis. Thus, it is possible to contemplate aspects in the understanding of the social production of ES and the biophysical valuation of ES as changes are projected to be evaluated. Confronted with a scenario of change, understanding the alteration of these relationships is relevant in an analysis for decision making. These evaluations involve aspects related to social assessment and analysis of trade-offs for the development of alternatives, and thus avoid environmental injustices, optimising the equitable distribution of ES and development opportunities.

9 Keys for Ecosystem Services Management for Inequality Alleviation

Different interdisciplinary conceptual frameworks are emerging, which analyse and respond to the growing socio-environmental conflicts. In terms of territory, ES, as well as environmental damage, are stratified, thus affecting the generation of spatial imbalances concerning those who benefit from the ecosystems and those who suffer their degradation. Unequal access to ES constitutes one of the main environmental problems, a source of ecological–distributive conflict in a society organised around scarce resources. Processes with objectives to justify and perpetuate political, social and economic arrangements (status quo) reinforce this relationship. Likewise, there is a clear link between a lower capacity for use of and access to ES, and poverty. Processes of social and environmental vulnerability, which foment poverty traps, strengthen this situation, thus generating unequal impacts depending on the geographical context considered. Moreover, the lack of agreed principles for decision making at different temporal and spatial scales, as well as the lack of integrated criteria between the evaluation of ES and the management thereof, have been serious limitations to the efforts made to apply the ES framework to decision making.

Addressing all these issues from a holistic perspective with thematic complexity is a challenge both to researchers and to decision makers. There is an urgent need to articulate these various theoretical visions and to adapt them to environmental management. In this way, it is key that decision making does not operate as a generator or aggravator of situations of poverty and vulnerability. To achieve this, it is crucial to conduct further research on these issues from an integrated perspective through interdisciplinary or transdisciplinary pathways. It is also necessary to involve stakeholders in these processes, through participatory and collaborative methodologies that develop and make explicit the different standpoints, values, interests, conflicts and power asymmetries. While those who manage and decide on ecosystems and biodiversity politically mediate the biophysical processes that condition the production of ES, they must then incorporate aspects of the social production of ES into management. Thus, this ES-based environmental management must differentiate the restrictions on the availability of ES from the underlying mechanisms that restrict access, in order to understand precisely the contributions of ES to human well-being and to the existing social stratification. Another priority should be the development of policies and investments that focus on the reduction of asymmetries that lead to a lack of opportunities (work, education, health and efficient access to ES).

The equitable distribution of ES, and its link to well-being and human relations, is still a relatively recent dimension for environmental justice. This approach to environmental conflicts from an ecological–distributive perspective is becoming a key element in decision making for the alleviation of inequalities. The successful integration of the environmental justice approach into environmental management based on ES depends on incorporating three fundamental dimensions that compromise it: first, building fair distributional scenarios between different social groups; second, the participation of social groups as a condition for the success of management scenarios; and third, the recognition of cultural factors that influence the production and social distribution of ES (Chaudhary *et al.*, 2018; Schlosberg, 2013).

The normative positioning of the ESPA framework, as a promoter of social equity and environmental sustainability, highlights its potential in conjunction with the environmental justice framework based on ES. This implies the generation of disaggregated ES analyses as well as the potential to avoid injustices derived from trade-offs and to include distributive dimensions of ES in the decision-making process.

In this review, as a response to the multiple conceptual frameworks and integration of approaches, the integral assessment of ES and biodiversity arises as a methodological proposal. This allows an environmental justice perspective in the analysis, by including sectors excluded from access to ES and incorporating notions of equity and vulnerability. The integration into social practice of value chains generated around ES and how they explain the value of ecosystem phenomena that support ES production contributes to the generation of valuable information concerning the expectations and pretensions of the social universe in an area to be managed. The implementation of these proposals involves the recognition of the value associated with biodiversity and ES as multidimensional, which lead to an effort to integrate different value schemes. The different multidisciplinary contributions dealt with transversally in this review can be valuable inputs to existing management tools such as SEAs and EIAs. Biased management can deepen the distributive gaps in the use of and access to ES. Addressing all these issues from a holistic perspective with thematic complexity is a challenge for both research and decision making. To respond to the growing socio-environmental conflicts, it is crucial to incorporate the dimension of social production of ES, as well as the mechanisms, which ultimately determine the distribution of well-being. However, these studies remain scarce today.

References

Adams, B. (2008) *Green Development: Environment and Sustainability in a Developing World* (3rd edn). Routledge. See https://doi.org/10.4324/9780203929711

Alberti, M. (2005) The effects of urban patterns on ecosystem function. *International Regional Science Review* 28 (2), 168–192. See https://doi.org/10.1177/0160017605275160

Alberti, M. (2008) *Advances in Urban Ecology: Integrating Humans and Ecological Processes in Urban Ecosystems*. New York: Springer.

Andersson, E., Barthel, S. and Ahrné, K. (2007) Measuring social-ecological dynamics behind the generation of ecosystem services. *Ecological Applications* 17 (5), 1267–1278.

Apollo, M. (2020) Human-free earth: The nearest future, or a fantasy? A lesson from artists. *International Journal of Environmental Studies* 77 (2), 360–366.

Armsworth, P.R., Chan, K.M.A., Daily, G.C., Ehrlich, P.R., Kremen, C., Ricketts, T.H. and Sanjayan, M.A. (2007) Ecosystem-service science and the way forward for conservation. *Conservation Biology* 21 (6), 1383–1384.

Barbier, E.B. (2010) Poverty, development, and environment. *Environment and Development Economics* 15 (6), 635–660.

Berkes, F., Folke, C. and Colding, J. (1998) *Linking Social and Biophysical Systems: Management Practices and Social Mechanisms for Building Resilience*. Cambridge: Cambridge University Press.

Berkes, F., Colding, J. and Folke, C. (eds) (2008) *Navigating Social-ecological Systems: Building Resilience for Complexity and Change* (pp. 31–189). Cambridge: Cambridge University Press.

Bharucha, Z. and Pretty, J. (2010) The roles and values of wild foods in agricultural systems. *Philosophical Transactions of the Royal Society B: Biological Sciences* 365 (1554), 2913–2926.

Bohman, J. (2007) Beyond distributive justice and struggles for recognition: Freedom, democracy, and critical theory. *European Journal of Political Theory* 6 (3), 267–276.

Bonds, M.H., Keenan, D.C., Rohani, P. and Sachs, J.D. (2009) Poverty trap formed by the ecology of infectious diseases. *Proceedings of the Royal Society B: Biological Sciences* 277 (1685), 1185–1192.

Boyce, J.K. (1994) Investing in human and natural capital. In A. Jansson, M. Hammer, C. Folke and R. Costanza (eds) *Investing in Natural Capital: The Ecological Economics Approach to Sustainability* (pp. 479–491). Washington, DC: Island Press.

Boyce, J.K. (2007) Is inequality bad for the environment? In R.C. Wilkinson and W.R. Freudenburg (eds) *Equity and the Environment (Research in Social Problems and Public Policy, Vol. 15)* (pp. 267–288). Bingley: Emerald Group Publishing Limited. https://doi.org/10.1016/S0196-1152(07)15008-0

Brondízio, E., Gatzweiler, F., Zografos, C., Kumar, M., Jianchu, X., McNeely, J., Kadekodi, G. and Martinez-Alier, J. (2010) Socio-cultural context of ecosystem and biodiversity valuation. In P. Kumar (ed.) *The Economics of Ecosystems and Biodiversity* (pp. 149–182). London and Washington, DC: Earthscan.

Brown, K. and Westaway, E. (2011) Agency, capacity, and resilience to environmental change: Lessons from human development, well-being, and disasters. *Annual Review of Environment and Resources* 36, 321–342.

Brown, T.C., Bergstrom, J.C. and Loomis, J.B. (2006) *Ecosystem Goods and Services – Definition, Valuation, and Provision*. Fort Collins, CO: Rocky Mountain Research Station, US Forest Service. See http://www.fs.fed.us/rm/value/docs/ecosystem_goods_services.pdf (accessed 10 December 2019).

Bullard, R.D. (1990) *Dumping in Dixie: Race, Class, and Environmental Quality*. Boulder, CO: Westview Press.

Butler, C. and Oluoch-Kosura, W. (2006) Linking future ecosystem services and future human well-being. *Ecology and Society* 11 (1), 30.

Butler, C., Chambers, R., Chopra, K., Dasgupta, P., Duraiappah, A.K., Kumar, P. and Niu, W.Y. (2003) Ecosystems and human well-being. In J. Alcamo *et al.* (eds) *Ecosystems and Human Well-being: A Framework for Assessment*. Washington, DC: Island Press.

Byrne, J., Martinez, C. and Glover, L. (2002) A brief on environmental justice. In J. Byrne, L. Glover and C. Martinez (eds) *Environmental Justice: Discourses in International Political Economy* (pp. 3–17). New Brunswick, NJ and London: Transaction Books.

Carpenter, S., Bennett, E. and Peterson, G. (2006) Scenarios for ecosystem services: An overview. *Ecology and Society* 11 (1), 29.

Carpenter, S.R., Mooney, H.A., Agard, J., Capistrano, D., DeFries, R.S., Díaz, S. and Perrings, C. (2009) Science for managing ecosystem services: Beyond the Millennium Ecosystem Assessment. *Proceedings of the National Academy of Sciences* 106 (5), 1305–1312.

Chan, K.M.A. and Satterfield, T. (2013) Justice, equity, and biodiversity. In S. Levin (ed.) *The Encyclopedia of Biodiversity* (pp. 434–441). Oxford: Elsevier.

Chan, K.M., Balvanera, P., Benessaiah, K., Chapman, M., Díaz, S., Gómez-Baggethun, E. ... and Turner, N. (2016) Opinion: Why protect nature? Rethinking values and the environment. *Proceedings of the National Academy of Sciences* 113 (6), 1462–1465.

Chaudhary, S., McGregor, A., Houston, D. and Chettri, N. (2018) Environmental justice and ecosystem services: A disaggregated analysis of community access to forest benefits in Nepal. *Ecosystem Services* 29, 99–115.

Costanza, R. (2008) Stewardship for a 'full' world. *Current History* 107 (705), 30–35.

Cote, M. and Nightingale, A.J. (2011) Resilience thinking meets social theory: Situating social change in socio-ecological systems (SES) research. *Progress in Human Geography* 36 (4), 475–489.

Cowling, R.M., Egoh, B., Knight, *et al.* (2008) An operational model for mainstreaming ecosystem services for implementation. *Proceedings of the National Academy of Sciences* 105 (28), 9483–9488.

Daily, G. (1997) *Nature's Services*. Washington, DC: Island Press.

Daw, T., Brown, K., Rosendo, S. and Pomeroy, R. (2011) Applying the ecosystem services concept to poverty alleviation: The need to disaggregate human well-being. *Environmental Conservation* 38 (4), 370–379. doi:10.1017/S0376892911000506

de Groot, R.S., Alkemade, R., Braat, L., Hein, L. and Willemen, L. (2010) Challenges in integrating the concept of ecosystem services and values in landscape planning, management, and decision making. *Ecological Complexity* 7 (3), 260–272.

Enfors, E.I. and Gordon, L.J. (2008) Dealing with drought: The challenge of using water system technologies to break dryland poverty traps. *Global Environmental Change* 18 (4), 607–616.

Ernstson, H. (2008) The social production of ecosystem services: Lessons from urban resilience research. In Rhizomia: Actors, networks, and resilience in urban landscapes. PhD thesis, Stockholm University.

Ernstson, H. (2013) The social production of ecosystem services: A framework for studying environmental justice and ecological complexity in urbanized landscapes. *Landscape and Urban Planning* 109 (1), 7–17.

Ernstson, H. and Sörlin, S. (2009) Weaving protective stories: Connective practices to articulate holistic values in the Stockholm National Urban Park. *Environment and Planning A* 41 (6), 1460–1479.

Escobar, A. (1999) Antropología y desarrollo. *Maguaré* 14, 42–73.

Escobar, A. (2006) Difference and conflict in the struggle over natural resources: A political ecology framework. *Development* 49 (3), 6–13. doi:10.1057/palgrave.development.1100267

Ferraro, P.J. and Hanauer, M.M. (2014) Quantifying causal mechanisms to determine how protected areas affect poverty through changes in ecosystem services and infrastructure. *Proceedings of the National Academy of Sciences* 111 (11), 4332–4337.

Fisher, B., Turner, R.K. and Morling, P. (2009) Defining and classifying ecosystem services for decision making. *Ecological Economics* 68 (3), 643–653.

Fisher, J.A., Patenaude, G., Meir, P., Nightingale, A.J., Rounsevell, M.D., Williams, M. and Woodhouse, I.H. (2013) Strengthening conceptual foundations: Analyzing frameworks for ecosystem services and poverty alleviation research. *Global Environmental Change* 23 (5), 1098–1111.

Fisher, J.A., Patenaude, G., Giri, K., Lewis, K., Meir, P., Pinho, P. and Williams, M. (2014) Understanding the relationships between ecosystem services and poverty alleviation: A conceptual framework. *Ecosystem Services* 7, 34–45.

Foladori, G. (2002) Avances y límites de la sustentabilidad social. *Economía, Sociedad y Territorio*, III (12), 621–637.

Forsyth, T. and Sikor, T. (2013) Forests, development and the globalization of justice. *The Geographical Journal* 179 (2), 114–121.

Ghermandi, A., Ding, H. and Nunes, P.A. (2013) The social dimension of biodiversity policy in the European Union: Valuing the benefits to vulnerable communities. *Environmental Science & Policy* 33, 196–208.

Gustavsson, M., Lindström, L., Jiddawi, N.S. and de la Torre-Castro, M. (2014) Procedural and distributive justice in a community-based managed marine protected area in Zanzibar. *Tanzania. Marine Policy* 46, 91–100.

Harvey, D. and Braun, B. (1996) *Justice, Nature and the Geography of Difference*. Oxford: Blackwell.

Janssen, M. and Ostrom, E. (2006) Empirically based agent-based models. *Ecology and Society* 11 (2), 37.

Jobbágy, E.G., Paruelo, J.M. and Laterra, P. (2011) *Valoración de servicios ecosistémicos: Conceptos, herramientas y aplicaciones para el ordenamiento territorial*. Buenos Aires: Ediciones INTA.

Jost, J. (2010) System justification theory. In J.M. Levine and M.A. Hogg (eds) *Encyclopedia of Group Processes and Intergroup Relations* (pp. 889–891). Thousand Oaks, CA: Sage.

Lade, S.J., Haider, L.J., Engström, G. and Schlüter, M. (2017) Resilience offers escape from trapped thinking on poverty alleviation. *Science Advances* 3 (5), e1603043. doi:10.1126/sciadv.1603043

Lakerveld, R.P., Lele, S., Crane, T.A., Fortuin, K.P.J. and Springate-Baginski, O. (2015) The social distribution of provisioning forest ecosystem services: Evidence and insights from Odisha, India. *Ecosystem Services* 14, 56–66.

Laterra, P., Jobbágy, E.G. and Paruelo, J.M. (2011). Valoración de servicios ecosistÚmicos. Conceptos, herramientas y aplicaciones para el ordenamiento territorial (No. P01 INTA 18477). INTA

Lele, S. (2013) *Environmentalists, Justices and the Limits of Ecosystem Services Frameworks*. London: Routledge.
Low, N. and Gleeson, B. (1998) Situating justice in the environment: The case of BHP at the OK Tedi Copper Mine. *Antipode* 30 (3), 201–226.
Martin, A., Gross-Camp, N., Kebede, B., McGuire, S. and Munyarukaza, J. (2014) Whose environmental justice? Exploring local and global perspectives in payments for ecosystem services scheme in Rwanda. *Geoforum* 54, 167–177.
Martínez Alier, J. (1995) De la economía ecológica al ecologismo popular (No. 333.72 M385 1995). Nordan Comunidad.
Martinez-Alier, J. (1995) The environment as a luxury good or 'too poor to be green'? *Ecological Economics* 13 (1), 1–10.
Martinez-Alier, J. (2002) Los indicadores de insustentabilidad urbana como indicadores de conflicto social. *Ayer* 46, 43–62.
Martinez-Alier, J. (2003) *The Environmentalism of the Poor: A Study of Ecological Conflicts and Valuation*. Cheltenham: Edward Elgar.
Martinez-Alier, J. (2005) Ecological economics as human ecology. In N. Chandra Sahu and A. Kumari Choudhury (eds) *Dimensions of Environmental and Ecological Economics* (pp. 47–80). Hyderabad: Universities Press.
Martínez-Alier, J. (2007) El ecologismo popular. *Revista Ecosistemas* 16 (3), 148–151.
Martínez-Alier, J.M. (2008) Conflictos ecológicos y justicia ambiental. *Papeles de Relaciones Ecosociales y Cambio Global* 103, 11–27.
Martínez Alier, J. (2009) *El ecologismo de los pobres: conflictos ambientales y lenguajes de valoración*. Barcelona: Icaria.
Martinez-Alier, J. (2012) Environmental justice and economic degrowth: An alliance between two movements. *Capitalism Nature Socialism* 23 (1), 51–73.
Martín-López, B. and Montes, C. (2010) Funciones y servicios de los ecosistemas: Una herramienta para la gestión de los espacios naturales. In *Guía Científica de Urdaibai*. Madrid: UNESCO, Dirección de Biodiversidad y Participación Ambiental del Gobierno Vasco.
Martín-López, B., González, J.A. and Vilardy, S. (2012) *Guía docente ciencias de la sostenibilidad*. Bogotá: CO-BAC.
Marzluff, J.M., Shulenberger, E., Endlicher, W., Alberti, M., Bradley, G., Ryan, C. and ZumBrunnen, C. (2008) *An International Perspective on the Interaction between Humans and Nature*. New York: Springer Science and Business Media.
McSweeney, K. (2005) Natural insurance, forest access, and compounded misfortune: Forest resources in smallholder coping strategies before and after Hurricane Mitch, northeastern Honduras. *World Development* 33 (9), 1453–1471.
MEA (Millennium Ecosystem Assessment) (2003) *Ecosystems and Human Well-being*. Washington, DC: Island Press.
MEA (Millennium Ecosystem Assessment) (2005) *Ecosystems and Human Well-being: A Synthesis*. Washington, DC: Island Press.
Mitchell, G. and Norman, P. (2012) Longitudinal environmental justice analysis: Co-evolution of environmental quality and deprivation in England, 1960–2007. *Geoforum* 43 (1), 44–57.
Müller, F., Baessler, C., Schubert, H. and Klotz, S. (2010) *Long-term Ecological Research*. Dordrecht: Springer.
Mullin, K., Mitchell, G., Nawaz, N.R. and Waters, R.D. (2018) Natural capital and the poor in England: Towards an environmental justice analysis of ecosystem services in a high-income country. *Landscape and Urban Planning* 176, 10–21.
Narayan, D., Chambers, R., Shah, M.K. and Petesch, P. (1999) *Global Synthesis: Consultations with the Poor*. Washington, DC: World Bank.
Norgaard, R.B. (2010) Ecosystem services: From eye-opening metaphor to complexity blinder. *Ecological Economics* 69 (6), 1219–1227.
Partidário, M. and Jesus, J. (2003) *Fundamentos de avaliação de impacte ambiental*. Lisbon: Universidade Aberta.

Paumgarten, F., Locatelli, B. and Witkowski, E.T.F. (2018) Wild foods: Safety net or a poverty trap? A South African case study. *Human Ecology* 46 (2), 183–195.

Pickett, S.T., Cadenasso, M.L., Grove, J.M., Groffman, P.M., Band, L.E., Boone, C.G. and Law, N.L. (2008) Beyond urban legends: An emerging framework of urban ecology, as illustrated by the Baltimore Ecosystem Study. *BioScience* 58 (2), 139–150.

Prato, T. (1999) Multiple attribute decision analysis for ecosystem management. *Ecological Economics* 30 (2), 207–222.

Ribot, J.C. and Peluso, N.L. (2003) A theory of access. *Rural Sociology* 68 (2), 153–181.

Rincón-Ruíz, A., Echeverry Duque, M.A., Piñeros Quiceno, A.M., Tapia Caicedo, C., Drews, A.D., Arias Arévalo, P. and Zuluaga Guerra, P.A. (2014) *Valoración integral de la biodiversidad y sus servicios ecosistémicos: Aspectos conceptuales y metodológicos.* Colombia: Instituto de Investigación de Recursos Biológicos Alexander von Humboldt.

Robards, M.D., Schoon, M.L., Meek, C.L. and Engle, N.L. (2011) The importance of social drivers in the resilient provision of ecosystem services. *Global Environmental Change* 21 (2), 522–529.

Rodríguez, J.P., Beard Jr, T.D., Bennett, E.M. *et al.* (2006) Trade-offs across space, time, and ecosystem services. *Ecology and Society* 11 (1), 28.

Sánchez-Moreno, S. and Ferris, H. (2007) Suppressive service of the soil food web: Effects of environmental management. *Agriculture, Ecosystems & Environment* 119 (1–2), 75–87.

Santos, C. (2010) Agua en Uruguay: Lucha social y la emergencia de nuevos esquemas de politización. *Theomai* 22, 76–85.

Santos, C. (2011) *¿Qué protegen las áreas protegidas?: Conservación, producción, Estado y sociedad en la implementación del Sistema Nacional de Áreas Protegidas.* Montevideo: Ediciones Trilce.

Scheffer, M., van Bavel, B., van de Leemput, I.A. and van Nes, E.H. (2017) Inequality in nature and society. *Proceedings of the National Academy of Sciences* 114 (50), 13154–13157.

Scherr, S.J. (2000) A downward spiral? Research evidence on the relationship between poverty and natural resource degradation. *Food Policy* 25 (4), 479–498.

Schlosberg, D. (2004) Reconceiving environmental justice: Global movements and political theories. *Environmental Politics* 13 (3), 517–540.

Schlosberg, D. (2007) *Defining Environmental Justice: Theories, Movements, and Nature.* Oxford: Oxford University Press.

Schlosberg, D. (2013) Theorizing environmental justice: The expanding sphere of a discourse. *Environmental Politics* 22 (1), 37–55. doi:10.1080/09644016.2013.755387

Seppelt, R., Dormann, C.F., Eppink, F.V., Lautenbach, S. and Schmidt, S. (2011) A quantitative review of ecosystem service studies: Approaches, shortcomings and the road ahead. *Journal of Applied Ecology* 48 (3), 630–636.

Sikor, T. (2013) The justices and injustices of ecosystem services. In T. Sikor (ed.) *The Justices and Injustices of Ecosystem Services* (pp. 24–28). London: Routledge.

Sikor, T., Martin, A., Fisher, J. and He, J. (2014) Toward an empirical analysis of justice in ecosystem governance. *Conservation Letters* 7 (6), 524–532.

Sörlin, S. (1998) Monument and memory: Landscape imagery and the articulation of territory. *Worldviews: Environment, Culture, Religion* 2, 269–279.

Stiglitz, J.E. (2012) *The Price of Inequality: How Today's Divided Society Endangers Our Future.* New York: WW Norton & Company.

Stiglitz, J.E. (2015) The price of inequality: How today's divided society endangers our future. In *Sustainable Humanity, Sustainable Nature: Our Responsibility* (pp. 379–399). New York: W.W. Norton & Company.

Tallis, H., Pagiola, S., Zhang, W., Shaikh, S., Nelson, E., Stanton, C. and Shyamsundar, P. (2011) Poverty and the distribution of ecosystem services. In P. Kareiva, H. Tallis, T.H. Ricketts, G.C. Daily and S. Polasky (eds) *Natural Capital: Theory and Practice of Mapping Ecosystem Services* (pp. 278–293). Oxford: Oxford University Press.

Tapella, E. (2012) Heterogeneidad social y valoración diferencial de servicios ecosistémicos. Un abordaje multi-actoral en el Oeste de Córdoba (Argentina). Dissertation, Universidad Nacional de Córdoba.

Tschakert, P. (2016) *The Role of Inequality in Climate-poverty Debates*. Washington, DC: World Bank.

Valpreda, C., Gudiño, M. and de Lillo, M.V. (2003) La evaluación de impacto ambiental y el Ordenamiento Territorial. *Sociologia Ruralis* 46 (1), 61–82.

Vilardy, S. and González, J.A. (2011) *Repensando la Ciénaga: Nuevas miradas y estrategias para la sostenibilidad en la Ciénaga Grande de Santa Marta*. Santa Marta: Universidad del Magdalena and Universidad Autónoma de Madrid.

Walker, B., Holling, C.S., Carpenter, S. and Kinzig, A. (2004) Resilience, adaptability, and transformability in social-ecological systems. *Ecology and Society* 9 (2), 5.

Wolch, J.R., Byrne, J. and Newell, J.P. (2014) Urban green space, public health, and environmental justice: The challenge of making cities 'just green enough'. *Landscape and Urban Planning* 125, 234–244.

2 The Climate Change–Migration Nexus Seen Through the Lens of the European Union: Analysis of Legal and Policy Frames

Agnieszka Legut

In recent years, the climate change–migration nexus has emerged as a salient issue within academia and also – to some extent – within legal and political debates at the international level. A number of terms have been used in an attempt to capture the complex phenomenon, each with its legal consequences and political weight. As of now, the debates result in only a few international political developments and a persistent protection gap in international law, regarding especially those who are forced to migrate across borders in connection to the effects of climate change. This chapter briefly presents the few major international documents that refer to the phenomenon of climate change induced migration. By reviewing the literature, it maps as well the terminology currently in use. Yet it focuses on the analysis of if and how the climate change–migration nexus is reflected within the legal and policy framework of the European Union (EU) – an important global actor and destination point on a number of South–North migration routes. In other words, by looking at policy documents and legal acts, the study aims to trace EU legal and policy frame(s) that have been developed in reference to the phenomenon of climate-induced migration. The analysis shows that there has been no comprehensive policy development at the EU level that would respond to the phenomenon in question, but rather that what can be traced are references scattered across a number of policy fields.

1 Introduction

It is comparatively recently that human mobility has become problematised in the context of climate change and has emerged as a topical issue within academia as well as international political and legal debates. In the process, a vast array of conceptual tools have been created in order to grasp the complexities of the climate change–migration nexus. The resulting rich yet debatable terminology comprises words and expressions with diverse legal and political implications. With the growing number of empirical studies available, it remains clear that knowledge of the phenomenon's scope and characteristics requires expanding further. Outside academia, the issue has received some attention which, however, has been translated into just a few international political developments. There persists a protection gap in international law, particularly with reference to those migrants who are forced to migrate across borders in relation to climate change impacts.

This chapter presents a brief outline of a current regional response to the phenomenon of climate-induced migration, i.e. that of the EU, which can be regarded an important actor as far as the global migration governance agenda is concerned (Geddes & Somerville, 2012). In particular, it describes *if* and *how* the climate change–migration nexus manifests itself in the legal and policy framework of the EU. By means of analysing policy documents and legal acts, the study aims to trace EU legal and policy frame(s) that have been developed in this respect.

2 Background

Humans have long been known to move once their area of living becomes 'environmentally or climatically inhospitable' (Maretti *et al.*, 2019: 142). In academic thought, changes in the natural environment have been seen as a significant trigger of human mobility (Piguet, 2013). It was, however, only comparatively recently that migration started to be considered in the context of changing environmental conditions resulting from anthropogenic impacts on climate. The climate change–migration nexus emerged in the second half of the 1980s in reports published by the United Nations Environmental Programme (UNEP), the Worldwatch Institute, the Intergovernmental Panel on Climate Change and a growing body of academic works (Maretti *et al.*, 2019; Piguet, 2013). Since then, and especially since the mid-2000s, it has become a topical issue within academia as well as a political activity of international institutions and advocacy groups (McAdam, 2011).

Within the debate, two major opposing positions have emerged, namely those of maximalists and minimalists (Surkhe, 1994), otherwise known as alarmists and sceptics (Maretti *et al.*, 2019) or more recently as proponents and critics (Morrissey, 2012), the main point of contention being whether there is a link between climate change and human mobility that could be conceptualised as direct and causal or if the nexus involves more complex and nuanced interconnections. Although there are still a number of understudied and unresearched areas in this respect (Brown, 2008; Castles, 2011; IOM, 2019; Kniveton *et al.*, 2008; Martin, 2010; Mence & Parrinder, 2017), more and more empirical analyses have revealed the complexity of the phenomenon (Maretti *et al.*, 2019). Migration and environmental scholars have arrived at a broad consensus that the link between climate change and mobility is 'complex, also largely unpredictable' (Brown, 2008: 12) and 'rarely linear' (IOM, 2019: 253); in other words, the former has 'a diffuse and indirect impact'(Mayer, 2018: 95) on the latter. In a nutshell, as McAdam (2011) pointed out:

> it is inherently fraught to speak of 'climate change' as the 'cause' of human movement, even though its impacts may exacerbate existing socio-economic or environmental vulnerabilities. Rather, climate change will have an 'incremental impact', 'adding to existing problems' and 'compounding existing threats'. (McAdam, 2011: 9)

The climate change–migration nexus is nowadays conceptualised as embedded in a context-dependent complex interplay of a wide range of factors – whether political, economic, social or demographic (Boas *et al.*, 2019; IOM, 2014; Martin, 2013; Mayer, 2018). Climate-induced movement is thus widely perceived as a multi-causal phenomenon. Studies carried out so far have indicated that it is often highly challenging, if not impossible, to isolate the drivers related to climate change from other triggers (Baldwin, 2017; Merone & Tait, 2018). Even in the seemingly straightforward case of the sinking islands that are threatened by the rising sea-level, attributing outward migration patterns only to climate change impacts is in fact an oversimplification. It has been reported that on Kiribati or Tuvalu, consequences of climate change act as a 'threat-multiplier' (Mence & Parrinder, 2017) or 'a tipping point' (McAdam, 2011) in exacerbating pre-existing socioeconomic issues such as, for instance, overcrowding or unemployment, and cannot be claimed to be a sole driver of out-migration (Hartmann, 2010: 238). Research has shown that climate-induced migration is also multidimensional. In response to a set of environmental and other stressors, mobility can develop into diverse patterns, starting from forced displacement and evolving into gradual outflow that is more voluntary in nature. It can involve crossing state borders or remain internal. The displacement can also vary in duration.

Because of its multi-causal and multidimensional nature, mobility related to climate change cannot be easily predicted as far as its volume and scope are concerned. Still, Norman Myers (1995) provided his estimates which, although later widely quoted, are rather 'guesstimates' – to quote a term coined by Kolmannskog (2008); in fact Myers himself declared they were based on 'heroic extrapolations' (Hartmann, 2010).

That is not to say, however, that migrant populations will not, grow in numbers[1] especially s ince with constantly rising average temperatures, some of the most adverse consequence may still lie ahead (Fornalé & Doebbler, 2017). That is to say, the study of complexities behind the climate change–migration nexus is still far from being comprehensively understood. Many agree that further research is required (Brown, 2008; Castles, 2011; IOM, 2019; Kniveton *et al.*, 2008; Martin, 2010; Mence & Parrinder, 2017). In particular, it should focus not on whether but on how climate change affects patterns of human migration, along with a wide range of other determinants (Boas *et al.*, 2019; Mayer, 2018).

2.1 Conceptual mapping

As demonstrated in Table 2.1, numerous terms have been used to refer to the complex phenomenon of climate-induced migration. The definitional debate, which is still ongoing (IOM, 2014; Ionesco & Chazalnoel, 2018), regards the theoretical and methodological grounding of the terms as well as their legal and political implications, since definitions are not mere semantic choices but are also 'crucial in guiding policies of governments and international agencies' (Castles, 2002: 9). What follows is a conceptual mapping exercise which aims to highlight the intricacies and implications of the terminology. It is presented along the lines of such key markers of human mobility as being (1) voluntary or involuntary and (2) internal or external. Both criteria are crucial for defining the responsibilities of nation-states and the international community towards those migrating.

Table 2.1 Terms used in reference to human mobility in the context of climate change

		Nature and scope of movement		
		Refugee	**Migrant**	**Displacee**
Nature and scope of environmental influence	Climate (change)	• Climate refugees • Forced climate migrants	• Climate (change) migrants • Climate (change)-induced migrants • Climate-related migration • Climate-driven migration	• Climate displacees • Climate change induced displacement • Disaster-displaced persons
		Climate mobilities		
	Environment	• Environmental refugees • Ecological refugees • Environmentally induced forced migrants	• Environmental migrants • Environmentally related migration	• Environmentally displaced persons

2.1.1 The environment versus climate change

When talking about environmental conditions and their influence on human mobility, a distinction is often made between slow- and rapid-onset phenomena, since they are likely to produce different patterns of migration. For instance, the typology introduced by (McLeman, personal communication) differentiates between climate processes and climate events, the former comprising slow-onset transformations (e.g. sea-level rise, desertification), the latter rapid-onset dramatic occurrences, frequently called natural disasters (e.g. floods, hurricanes) (Brown, 2008: 19). Climate events are often associated with a sudden forced displacement of a substantial number of people, yet a displacement that does not necessarily involve border crossing and is usually of a temporary character. On the other hand, climate processes may act as a threat multiplier of existing vulnerabilities and cause people to move in search of safer conditions and better opportunities. As such, it is linked to migration seen primarily as a result of unconstrained decision making. Yet the voluntary character of the case of slow-onset degradation can be questioned (Merone & Tait, 2018). Climate processes can also mean depleting livelihoods, which can hinder people's capacity to migrate (Geddes & Somerville, 2012).

Some of the terms that have been coined attempt to capture the multicausality of the phenomenon discussed. Migration is thus environmentally (or else climate change) 'induced' or 'related' rather than caused by it. Other descriptors point to a more determining impact of environmental factors, presenting migration as 'driven' or 'triggered' by either more generalised environmental or more specific climate change factors. The terms 'climate refugee' and 'climate migrant' may in turn suggest a direct dominant impact of climatic phenomena on mobility. Still, they have been criticised, taken the difficulty of isolating other drivers of migration, such as dire economic circumstances, from climate change impacts – as those tend to be intertwined.

2.1.2 Refugee versus migrant

Fundamentally, the conceptual difference between a 'refugee' and a 'migrant' is based on an evaluation as to whether the mobility is of a voluntary or non-voluntary character. The migrant – being also a more general term – is associated with 'a voluntary move towards a more attractive lifestyle' (Brown, 2008: 13), whereas the refugee indicates a person who has been forced to migrate. In practice, the distinction between the voluntary and involuntary nature of human mobility is often blurred. Rather, the two characteristics should more accurately be thought of as two extremes on a continuum (IOM, 2019). Still, dividing migrants into these two major categories, i.e. voluntary and forced, is one of the most important premises underlying migration policies and relevant international law. Refugees and other vulnerable categories of migrants are subject to the 'non-refoulement principle' which implies the responsibility of a third state to grant them temporary protection,[2] while migrants' right to enter and reside in a country is a subject to often restrictive immigration regimes.

The use of the terms 'migrant' and, in particular, 'refugee' in the context of climate-induced mobility has been surrounded by controversy. In the first place, a question has arisen as to their scope. Typically, it is the category of refugee that is associated with a right to protection by a third state. Should then also all those who are not displaced as the result of a sudden natural disaster but who migrate because of slow-onset degradation of their environment be included in the category of climate/environmental refugees? As mentioned above, their decision to move can hardly be seen as entirely voluntary. If yes, is there a need to set a benchmark for a 'tipping point' that renders a given area degraded enough to make migration away from it a forced move?

It is the term 'climate refugee', however, that has received particularly severe criticism – as a conceptually flawed legal misnomer (Kibreab, 1997; McGregor, 1993). Originally, with the emergence of interest in the climate change–migration nexus in the mid-1980s, 'environmental refugee' was used[3] and quickly gained popularity, especially within what came to be called the alarmist or maximalist approach. Later, some researchers stated that the term is conceptually misleading because of the implication (already mentioned above) that there exists a direct linear connection between climate change and forced migration. Also, climate refugees are generally not recognised either in the legal literature or in legal practice so far[4] as those that qualify for protection under the framework of the 1951 Refugee Convention (van der Vliet, 2018). The core international legal document for recognising the status of refugees adopts a narrow interpretation, defining a refugee as somebody who lives in a 'well-founded fear of being persecuted' for reasons that are enumerated to be 'race, religion, nationality, membership of a particular social group or political opinion' (Williams, 2008: 507).[5] Others have pointed out, however, that this legal category has also been subject to interpretation over time (Behrman & Kent, 2018; McAdam, 2009).

2.1.3 Displacee versus refugee

Traditional formal understanding of refugee status includes crossing state borders in search of third country protection, whereas a 'displaced person' or 'displacee' is associated with internal migration (Williams, 2008), although the word 'displacement' has been used in reference to mobility as such. It has been indicated that restricting terminology on climate change/environmental mobility to that of refugee excludes those who are displaced by climate change impacts, yet stay within their state borders (Brown, 2008). As shown by empirical research, often climate-induced migration does not involve trans-border movement (Brown, 2008; Martin, 2010; McAdam, 2011; Mence & Parrinder, 2017).

In spite of all the controversies, the term 'climate refugees' is still in use. Some have decided to employ it as an advocacy tool 'for highlighting the responsibilities of industrialized nations for global environmental degradation, in other words for promoting climate justice' (Gemenne, 2015). Others have chosen to introduce more nuanced descriptive terms, such as 'environmentally induced forced migrants' or 'forced climate migrants' (Anopama Atapattu, 2018).[6] A shift has also been observed towards more inclusive general terminology based on the word migrant rather than refugee (Bettini *et al.*, 2017). For instance, the International Organization for Migration has adopted the term 'environmental migrants'.[7] Boas *et al.* (2019) have suggested introducing an even wider term, i.e. 'climate mobilities', to encompass multiple migration patterns that can result from climate change impacts.

2.2 Existing legal frameworks and political responses

Taking into consideration a rather wide range of migratory reactions under climate change stressors, it can be stated that at least some categories of migrants require protection by their own or a third country. Currently, there is no legal framework that would address the issue comprehensively at the international level.[8] The existing measures are fragmented and mainly inadequate. In case of climate-induced forced migration that is trans-border in character, the narrow definition of the 1951 Refugee Convention renders the recognition of refugee status on the grounds of climate change displacement inapplicable. The provisions that provide complementary protection in cases that do not fall under the scope of the document (for instance, the obligation of non-refoulement[9] or human rights principles), although with some potential for future extensive interpretations (Perrini, 2018), are at present hardly valid for climate forced migrants. Williams concludes that:

> complementary protection offers little additional value to securing the protection of those displaced by environmental change. Operating largely on an ad hoc basis, and unable to recognize or address the underlying causes of environmental displacement, this regime is currently better suited to other refugee groups seeking alternative protection, rather than those subject to environmentally motivated displacement. (Williams, 2008: 514)

As far as persons internally displaced for environmental reasons are concerned, there exists a legal framework based on the 1998 *UN Guiding Principles on Internal Displacement*. The document is dedicated to people displaced as an effect of rapid-onset climate events or human-made disasters, and as such they can be seen as 'largely conducive to recognizing and supporting environmental refugees'(Williams, 2008: 512), especially once it was extended to include those climate forced migrants who in their displacement crossed the borders of their state. Still, the document has a legally non-binding status which limits its applicability (Williams, 2008).

Alongside the interest within academia in climate-induced migration, the issue also became topical in political debates at the international level. Its emergence under the IOM and its sustained presence on the organisation's agenda can be regarded as 'a barometer of states' interest in climate migration' and indicate that it is nowadays seen as 'a key migration policy question' (Ionesco & Chazalnoel, 2018: 107). Added to that, mobility in the context of climate change has been included in two important global policy processes[10] developed within the United Nations, i.e. that of the Framework Convention on Climate Change (UNFCCC) and the Global Compact for Safe, Orderly and Regular Migration. The latter document represents a serious step forward since it recognises that climate change can be an important trigger of mobility and confirms the multi-causal and multidimensional character of climate-induced migration (Ionesco & Chazalnoel, 2018). Also worth mentioning in this context is the Nansen Initiative, which is a multilateral consultative process, started in 2012 by the governments of Switzerland and Norway. It focuses on cross-border displacement due to natural disasters and its scope is limited to migration as a result of severe weather events. On the one hand, this makes it rather restrictive but, on the other, in those cases, establishing the causal link between displacement and the rapid onset event can be possible (Anopama Atapattu, 2018).

Even though in recent years there have clearly been some political developments with a focus on climate-induced migration, a dedicated comprehensive global governance framework has not yet emerged (IOM, 2019; Ionesco & Chazalnoel, 2018).

3 Methods

In order to analyse *if*, and possibly *how* climate-induced migration has become a regional focus within the EU, a set of analytical tools have been devised that were inspired by critical frame analysis and in particular by the works of Goffmann (1974), Entman (1993) and Verloo (2005). In general terms, frames are interpretative schemata which structure our understanding of reality (Goffmann, 1974). A policy frame can be defined as 'an organising principle that transforms fragmentary or incidental information into a structured and meaningful problem, in which a solution is implicitly or explicitly included' (Verloo, 2005: 20). As Entman (1993) suggests, such an organising principle consists of a set of elements that name and interpret the problem as well as the actors involved, define the problem's cause and indicate a possible solution to it. When an issue is not problematised, the fact can still be regarded as a meaningful discursive act. In other words, no policy is also a policy position (McNamara, 2007).

The documents for analysis have been selected under two major criteria: firstly, if they addressed the topic of climate change induced migration; secondly, if they fell into the category of policy document or legal act. Other publications such as, for instance, research reports or briefs prepared for political groups in the European Parliament, which would otherwise have been an interesting extension of the study, were excluded. The findings of the analysis, based on documents (see Annex for the full list) from a period of almost two decades (1999–2018), are presented here mostly from a synchronous perspective due to the compact nature of the paper.

4 Findings: EU Legal and Policy Frames on Climate-induced Migration

Within the legal framework of the EU, there is no provision dedicated to climate forced migrants. Art. 78 of TFUE[11] directly recognises the core international interpretation of refugee status, as it is set out in the 1951 Geneva Convention and the 1967 Protocol, as binding for EU Member States. Some authors interpret the existing complementary protection measures comprising the EU Qualification Directive and Temporary Protection Directive as offering – although only indirectly – some form of international protection for those forced to migrate due to environmental stressors. It needs to be emphasised, however, that none of the legal instruments mentions climate change forced migrants explicitly (de Moor & Cliquet, 2009; Perrini, 2018). Neither has it been given a mention in the draft proposal for a new Qualification Regulation (Jakobsson, 2019). Added to that, both directives are based on a political compromise between EU Member States and as such are founded on the lowest common denominator. What results is 'a diluted protection system' (Williams, 2008: 513).

Those climate-induced migrants whose decision to move is largely voluntary are also omitted by the EU legal framework. The immigration regimes are still mostly a matter of national competence[12] and, given the political context and high securitisation of immigration in public discourse(s), it is not likely to change in the immediate future so as to encompass legal migration programmes aimed at those who decide to migrate under environmental stressors.

While the EU legal frame on climate-induced migration seems to be one of *invisibility*, policy documents mention this type of human movement relatively often. However, there is no single comprehensive document dealing with this issue and the phenomenon is still mostly presented as *intangible*. The climate change–migration nexus is often mentioned in the context of the necessity for further debate and research. Also the choice of terminology used to refer to the phenomenon in question reflects a strong preference for descriptive, broad terms such as 'climate/environmentally induced displacement' ('environmental/climate displaced person') to imply forced mobility and 'climate-induced migration' ('climate-induced migrant') in reference to more voluntary movement. The rationale behind such a choice is indicated to be twofold: firstly, broader terms do not hinder further comprehensive study of the phenomenon and, secondly, stronger terms such as climate refugee would entail governance and protection responsibility. What is more, climate-induced migration is depicted in the analysed EU policy documents as intangible since it presents identification difficulties due to its multidimensional character and also because of the fact that, although climate change is declared to be the root cause of diverse migration reactions, the phenomenon is still perceived as multi-causal.

Climate-induced mobility is defined (especially in the earlier of the studied EU documents) as *a security problem* that can take two forms, i.e. that of the mass influx of migrants as well as that of a conflict trigger and multiplier. The first manifestation is perceived largely as a direct security challenge to the EU, while the other is seen mostly as a challenge to the stability of vulnerable third countries. Depending on which of the two parties involved is emphasised, different policy frames emerge.

At the beginning, when the estimates by Myers (1993) began to circulate, there was a conviction that, due to climate change impacts, mass displacement on an unprecedentedly large scale was imminent by 2050. It also found expression in the EU policy frame in which climate-induced migration is securitised. With more research available and a more nuanced approach to the phenomenon in question, as well as the current conviction that climate-induced migration is predicted to be in the most part internal, the problem started to be perceived to a greater degree as externalised. The likelihood of climate-related mass influx is assessed in the analysed documents as currently relatively low. Similarly, the priority of the issue of climate-related mobility is downplayed. Thus, climate-induced migration is constructed as a *development challenge* and potential *humanitarian issue* in response to rapid onset events. In consequence, the focus is not as much on the climate change–migration nexus as on the interconnection between climate change and development. Such a definition of the problem prompts a response in the form of a focus on building resilience capacities and assisting general development in the vulnerable third countries.

When still viewed from the perspective of the EU as the key involved actor and destination point for climate-related mobility, the phenomenon is seen in the context of *protection concern*. Here a recommended solution is mostly to elaborate on the existing legal provisions and to further develop the international framework in this respect.

5 Conclusion

The above brief outline of the EU legal and policy frames in reference to climate-induced migration confirms what other researchers have observed as far as the relevant EU stance is concerned. Even though the issue is present in policy discussions at the EU level, it is scattered among a number of policy fields such as, for instance, security, development, humanitarian response, migration and asylum. There is, in other words, no comprehensive programmatic answer that would address the phenomenon in its complexity (Jakobsson, 2019). Also, currently it seems to be a thematic field of low priority within the EU (Petrillo, 2015). Still, with temperatures rising, desertification and sea-level rise, to name but a few climate change manifestations, it seems crucial not to shift either political or academic attention from deliberating on the climate change–migration nexus. As Mence and Parrinder point out:

> The difficulties involved in conceptualising and defining the complexities of environmentally related migration are central to the challenge of increasing the knowledge base and engaging in effective policy deliberations and the formulation of policy responses. (Mence & Parrinder, 2017: 319)

Notes

(1) For more on estimating the volume of climate-induced migration, see Gemenne (2011).

(2) Formally speaking, the principle 'prohibits states from returning (...) an asylum seeker to territories where there is a risk that his or her life or freedom would be threatened on account of reasons mentioned in various human rights conventions' (van der Vliet, 2018: 22).

(3) The term 'environmental refugee' was popularised by El-Hinnawi (1985) of the United Nations Environment Programme (UNEP) (McAdam, 2011).

(4) McAdam (2011: 12–13) reports: 'So far, there have been a small number of cases in Australia and New Zealand where people from Tuvalu and Kiribati have sought to argue they should receive refugee protection from climate change impacts (...) They have all failed'.

(5) The 1951 Refugee Convention recognises a refugee to be a person who 'owing to a well-founded fear of being persecuted for reasons of race, religion, nationality, membership of a particular social group or political opinion, is outside the country of his nationality and is unable, or owing to such fear, is unwilling to avail himself of the protection of that country, or who, not having a nationality and being outside the country of his former habitual residence as a result of such events, is unable or, owing to such fear, is unwilling to return to it'.

(6) The author suggests the following definition of 'forced climate migrant': 'people who are forced to leave their homes or land either temporarily or permanently due to significant environmental damage associated with climate change or where their national state is no longer habitable. Where they have crossed an international border, the receiving state should accord basic human rights protection, until such time as they are able to return to their homes (in the case of temporary displacement) or in the case of permanent displacement, necessary legal arrangements should be made for their transition as residents of the receiving country with access to basic services and amenities' (Anopama Atapattu, 2018: 40).

(7) Environmental migrants in the IOM understanding mean: 'persons or groups of persons, who, for compelling reasons of sudden or progressive changes in the environment that adversely affect their lives or living conditions, are obliged to leave their habitual homes, or chose to do so, either temporarily or permanently, and who move either within their country or abroad' (IOM, 2007).

(8) There exist a few regional (the OAU Kampala Convention, Cartagena Declaration) and national (in Sweden or Finland) measures that can – usually indirectly – apply to some extent to climate forced migration (McAdam, 2011; Perrini, 2018; van der Vliet, 2018).
(9) The non-refoulement principle is set out, for example, in the Convention against Torture and Other Cruel, Inhumane or Degrading Treatment or Punishment of 1984 or the 1966 International Covenant on Civil and Political Rights.
(10) There are a number of other policy processes and initiatives with a focus on climate-induced migration. Their up-to-date list can be found in the IOM (2019) *World Migration Report 2020*.
(11) Article 78 of the Treaty on the Functioning of the European Union says:
(1) The Union shall develop a common policy on asylum, subsidiary protection and temporary protection with a view to offering appropriate status to any third-country national requiring international protection and ensuring compliance with the principle of non-refoulement. This policy must be in accordance with the Geneva Convention of 28 July 1951 and the Protocol of 31 January 1967 relating to the status of refugees, and other relevant treaties. (...). (Consolidated version of the Treaty on the Functioning of the European Union – Part Three: Union Policies and Internal Actions – Title V: Area of Freedom, Security and Justice – Chapter 2: Policies on border checks, asylum and immigration – Article 78 (ex Articles 63, points 1 and 2, and 64(2) TEC))
(12) Apart from policies on family reunification, long-term residents or highly skilled migration.

References

Anopama Atapattu, S. (2018) A new category of refugees? 'Climate refugees' and a gaping hole in international law. In S. Behrman and A. Kent (eds) *'Climate Refugee': Beyond the Legal Impasse?* (pp. 34–51). London: Routledge.

Baldwin, A. (2017) Climate change, migration, and the crisis of humanism. *WIREs Climate Change* 8, 1–7.

Behrman, S. and Kent, A. (eds) (2018) *'Climate Refugees': Beyond the Legal Impasse?* London: Routledge.

Bettini, G., Nash, S.L. and Gioli, G. (2017) One step forward, two steps back? The fading contours of (in)justice in competing discourses on climate migration. *The Geographical Journal* 183 (4), 348–358.

Boas, I., Farbotko, C. and Adams, H. (2019) Climate migration myths. *Nature Climate Change* 9, 901–903.

Brown, O. (2008) *Migration and Climate Change.* Geneva: International Organisation for Migration.

Castles, S. (2002) *Environmental Change and Forced Migration: Making Sense of the Debate.* Geneva: UNHCR Evaluation and Policy Analysis Unit.

Castles, S. (2011) Concluding remarks on the climate change–migration nexus. In E. Piguet, A. Pécoud and P. de Guchteneire (eds) *Migration and Climate Change* (pp. 415–27). Cambridge: Cambridge University Press.

de Moor, N. and Cliquet, A. (2009) Environmental displacement: A new challenge for European migration policy. *Proceedings of Protecting People in Conflict and Crisis: Responding to the Challenges of a Changing World.* Oxford: Refugee Studies Centre.

El-Hinnawi, E. (1985) *Environmental Refugees United Nations Environmental Program.* Nairobi.

Entman R. (1993) Framing: Towards clarification of a fractured paradigm. *Journal of Communication* 4, 51–58.

Fornalé, E. and Doebbler, C. (2017) UNHCR and protection and assistance for the victims of climate change. *The Geographical Journal* 183 (4), 329–335.

Geddes, A. and Somerville, W. (2012) Migration and environmental change in international governance: The case of the European Union. *Environment and Planning C: Government and Policy* 30 (6), 1015–1028.

Gemenne, F. (2011) Why the numbers don't add up: A review of estimates and predictions of people displaced by environmental changes. *Global Environmental Change* 21, 41–49.

Gemenne, F. (2015) One good reason to speak of 'climate refugees'. *Forced Migration Review* 49, 70–71.

Goffman, E. (1974) *Frame Analysis: An Essay on the Organization of Experience*. Boston, MA: Northeastern University Press.

Hartmann, B. (2010) Rethinking climate refugees and climate conflict: Rhetoric, reality and the politics of policy discourse. *Journal of International Development* 22 (2), 233–246.

IOM (International Organization for Migration) (2007) Discussion note: Migration and the environment, 94th session, MC/INF/288.

IOM (International Organization for Migration) (2014) *IOM Outlook on Migration, Environment and Climate Change*. Geneva: UN Migration.

IOM (2018) Global Compact for Safe, Orderly and Regular Migration. See https://www.un.org/en/ga/search/view_doc.asp?symbol=A/RES/73/195 (accessed 1 July 2019).

IOM (International Organization for Migration) (2019) *World Migration Report 2020*. Geneva: UN Migration.

Ionesco, D. and Chazalnoel, M.T. (2018) Advancing the global governance of climate migration through the United Nations Framework Convention on Climate Change and the Global Compact for Migration: Perspectives from the International Organization for Migration. In S. Behrman and A. Kent (eds) *'Climate Refugees': Beyond the Legal Impasse?* (pp. 103–117). London: Routledge.

Jakobsson, E. (2019) *Climate Change and Migration: Policy Approaches for a Sustainable Future*. ELF (European Legal Forum).

Kibreab, G. (1997) Environmental causes and impact of refugee movements: A critique of the current debate. *Disasters* 21 (1), 20–38.

Kniveton, D., Schmidt-Verkerk, K., Smith, C. and Black, R. (2008) *Climate Change and Migration: Improving Methodologies to Estimate Flows*. Geneva: International Organization for Migration.

Kolmannskog, V. (2008) *Future Floods of Refugees*. Oslo: Norwegian Refugee Council.

Maretti, M., Tontodimamma, A. and Biermann, P. (2019) Environmental and climate migrations: An overview of scientific literature using a bibliometric analysis. *International Review of Sociology* 29 (2), 142–158.

Martin, S. (2010) *Climate Change and International Migration*. Washington, DC: German Marshall Fund.

Martin, S.F. (2013) Environmental change and migration: What we know. *Migration Policy Institute Policy Brief* 2, 1–10.

Myers, N. (1993) Environmental refugees in a globally warmed world. *Bioscience* 43 (11), 752–761.

Mayer, B. (2018) Who are 'climate refugees'? Academic engagement in the post-truth era. In S. Behrman and A. Kent (eds) *'Climate Refugees': Beyond the Legal Impasse?* (pp. 89–100). London: Routledge.

McAdam, J. (2009) Environmental migration governance. *UNSW Law Research Paper* 1.

McAdam, J. (2011) *Climate Change Displacement and International Law: Complementary Protection Standards*. Geneva: UNHCR.

McGregor, J. (1993) Refugees and the environment. In R. Black and V. Robinson (eds) *Geography and Refugees: Patterns and Processes of Change* (pp. 157–170). London: Belhaven Press.

McNamara, K.E. (2007) Conceptualizing discourses on environmental refugees at the United Nations. *Population and Environment* 29 (1), 12–24.
Mence, V. and Parrinder, A. (2017) Environmentally related international migration: Policy challenges. In M. McAuliffe and K. Koser (eds) *A Long Way to Go: Irregular Migration Patterns, Processes, Drivers and Decision-making.* Canberra: ANU Press.
Merone, L. and Tait, T. (2018) 'Climate refugees': Is it time to legally acknowledge those displaced by climate disruption? *Australian and New Zealand Journal of Public Health* 42 (6), 508–509.
Morrissey, J. (2012) Rethinking 'the debate on environmental refugees': From 'maximalists and minimalists' to 'proponents and critics'. *Journal of Political Ecology* 19, 36–49.
Myers, N. (1995) *Environmental Exodus: An Emergent Crisis in the Global Arena.* Washington, DC: Climate Institute.
Myers, N. (2002) Environmental refugees: A growing phenomenon of the 21st century. *Philosophical Transactions: Biological Sciences* 357 (1420), 609–613.
Perrini, F. (2018) *Cambiamenti climatici e migrazioni forzate: Verso una tutela internazionale dei migranti ambientali.* Naples: Editoriale Scientifica.
Petrillo, E.R. (2015) Environmental migrations from conflict-affected countries: Focus on EU policy response. Working Paper No. 6. The Hague: The Hague Institute for Global Justice.
Piguet, E. (2013) From 'primitive migration' to 'climate refugees': The curious fate of the natural environment in migration studies. *Annals of the Association of American Geographers* 103 (1), 148–162.
Suhrke, A. (1994) Environmental degradation and population flows. *Journal of International Affairs* 47 (2), 473–496.
UN (1992) Framework Convention on Climate Change (UNFCCC). See https://treaties.un.org/pages/ViewDetailsIII.aspx?src=TREATY&mtdsg_no=XXVII-7&chapter=27&Temp=mtdsg3&clang=_en (accessed 1 July 2019).
UNHCR (1998) UN Guiding Principles on Internal Displacement. See https://www.unhcr.org/protection/idps/43ce1cff2/guiding-principles-internal-displacement.html (accessed 1 July 2019).
van der Vliet, J. (2018) 'Climate refugees': A legal mapping exercise. In S. Behrman and A. Kent (eds) *'Climate Refugees': Beyond the Legal Impasse?* (pp. 16–33). London: Routledge.
Verloo, M. (2005) Reflections on the concept and practice of the Council of Europe approach to gender mainstreaming. *Social Politics* 12 (3), 344–365.
Williams, A. (2008) Turning the tide: Recognizing climate change refugees in international law. *Law & Policy* 30 (4), 502–529.

Annex

Consolidated version of the Treaty on the Functioning of the European Union – Part 3: Union Policies and Internal Actions – Title V: Area of Freedom, Security and Justice – Chapter 2: Policies on border checks, asylum and immigration. See https://eur-lex.europa.eu/legal-content/EN/TXT/?uri=CELEX%3A12016E079 (accessed 2 April 2020).
Council Directive 2001/55/EC of 20 July 2001 on minimum standards for giving temporary protection in the event of a mass influx of displaced persons and on measures promoting a balance of efforts between Member States in receiving such persons and bearing the consequences thereof. See https://eur-lex.europa.eu/legal-content/EN/TXT/?uri=CELEX%3A32001L0055 (accessed 2 April 2020).
Directive 2011/95/EU of the European Parliament and of the Council of 13 December 2011 on standards for the qualification of third-country nationals or stateless persons as beneficiaries of international protection, for a uniform status for refugees or for persons eligible for subsidiary protection, and for the content of the protection granted.

See https://eur-lex.europa.eu/legal-content/EN/TXT/?uri=celex%3A32011L0095 (accessed 2 April 2020).

European Commission (2008) Climate change and international security: Paper from the High Representative and the European Commission to the European Council. See https://www.consilium.europa.eu/ueDocs/cms_Data/docs/pressdata/EN/reports/99387.pdf (accessed 2 April 2020).

European Commission (2011) 'Climate refugees'. Legal and policy responses to environmentally induced migration. See https://www.europarl.europa.eu/thinktank/en/document.html?reference=IPOL-LIBE_ET%282011%29462422 (accessed 2 April 2020).

European Commission (2012) Human rights and climate change: European Union policy options. See https://www.europarl.europa.eu/RegData/etudes/etudes/join/2012/457066/EXPO-JOIN_ET(2012)457066_EN.pdf (accessed 2 April 2020).

European Commission (2013) Commission Staff Working Document 'Adaptation to climate change impacts on human, animal and plant health', accompanying the document 'An EU Strategy on adaptation to climate change'. See https://ec.europa.eu/clima/sites/clima/files/adaptation/what/docs/swd_2013_136_en.pdf (accessed 2 April 2020).

European Commission (2015) European Agenda on Migration, COM (2015) 240 final. See https://ec.europa.eu/anti-trafficking/sites/antitrafficking/files/communication_on_the_european_agenda_on_migration_en.pdf (accessed 2 April 2020).

European Council (2010) *The Stockholm Programme: An Open and Secure Europe Serving and Protecting Citizens* 2010/C 115/01. Brussels: European Council.

European Parliament (1999) Resolution on the Environment, Security and Foreign Policy, passed on January 29, 1999, A4–0005/99, JO C 128/92 of 7 May 1999.

European Parliament (2009) White Paper 'Adapting to climate change: Towards a European framework for action', INI/2009/2152.

European Parliament (2017) Resolution on addressing refugee and migrant movements: The role of EU External Action (2015/2342 (INI)).

European Parliament (2018) Resolution on progress on the UN Global Compacts for Safe, Orderly and Regular Migration and on Refugees (2018/2642 (RSP)).

3 Dangers of Uncontrolled Urbanisation for Society: Selected Aspects

Małgorzata Kmak

The research issues concerning cities are very important and highly topical. The process of urbanisation has become an essential quality of modern civilisation. Urbanisation is a necessary process for both developed and developing countries. For the former, it is a natural consequence of change; for the latter – a stimulus for transformation. The nature of urbanisation is a reflection of the political, social and economic circumstances in which it occurs. The rapid urbanisation in developing countries is increasing social divides. This chapter aims to present an overview of the problems associated with global urbanisation and identifies: (1) the differences in the level of development on different continents; (2) the different ways in which urbanisation manifests itself; and (3) the negative consequences of urbanisation in the context of the problems occurring in the urban space such as homelessness, poverty and exclusion, and new solutions based on a compromise. Complementary local, regional and national policies in strategic areas of development are necessary. At the same time, the existing strategies and models of urban development are not effective enough to face the problems of 21st century cities. The instruments of exploitation of modern cities are ineffective in serving cities inhabited by millions of people. The slogan of sustainable development, so popular in planning documents, is not implemented in practice at all.

1 Introduction

Urbanisation is a necessary process for both developed and developing countries. Firstly, it is a natural consequence of change and, secondly, it is a stimulus for transformation. Uncontrolled urbanisation poses a serious social, economic and ecological threat. The formation of slums is one of the many consequences. New solutions are needed to tackle problems in the city space, such as homelessness, poverty and exclusion. Meanwhile, existing strategies and models of urban development are not effective enough to be able to meet the challenges arising in 21st century cities. The problem of uncontrolled urbanisation is multidimensional and rarely addressed in scientific discourse. Achieving prosperity has become the dominant goal in every society and region, regardless of the level of wealth already achieved and cultural and political traditions. The concept of a social city is controversial in social discourse, because it is often contrary to the vision of convenience and comfort. Admittedly, this is not the vision of the average inhabitant of a modern city; however, it is a vision of people whose decisions create business policy and define its principles and priorities. As a result, these decisions have a huge impact on the city's image and the living patterns of its inhabitants.

2 Background

Tell Qaramel in Syria, dating back to the 11th and 10th centuries BC (Kolińska, 2009), Çatalhöyük in Turkey (Hodder & Marciniak, 2015) or Jericho in Israel are considered the oldest cities in the world. Urbanism as an analytical category appeared in the 1930s. In his 1938 piece entitled 'Urbanism as a way of life', Louis Wirth described urbanism as a specific way of life that can be analysed empirically from three interrelated perspectives: (1) as a physical structure that encompasses the basic population, technology and ecological order; (2) as a system of social organisation [...]; and (3) as a collection of attitudes and ideas, as well as sets of personalities (Wirth, 1996). A city in the social sense, as Louis Wirth recreated it, is characterised by the diversity of people and urban space in the form of welfare zones and economic potentials as well as areas of poverty and material neglect. Diversity, as a necessity in the way urban space is organised, was recognised by Jane Jacobs (1961) in a study entitled *The Death and Life of Great American Cities*. According to the author, the function of cities and their streets is usefulness. They should not be shaped in a modernist way that is homogeneously and aesthetically simple. They should be full of visual surprises. In the author's opinion, higher buildings are superfluous because only rich users can bear the general costs of their maintenance. The exploitation of small blocks of flats is cheaper and allows a greater variety of forms and cubature. The author suggests three basic rules for increasing the feeling of safety: firstly, a clear division between the public and private space; secondly, the orientation of buildings on the street so that they are not facing side walls; thirdly, users appearing in the streets at different times of the day. A well-designed street and one that is used socially is a safe space, while a deprived area becomes increasingly dangerous. The achievement of diversity, which favours social inclusion, seems to pose the greatest challenge. The effectiveness of such a planned urban structure is dependent on, among other things, the organisation of basic functions and services (education, work, museums, schools and public buildings), low-rise building design, the use of old buildings and the occupational diversification of the inhabitants. Jacobs stresses that each district or estate should have at least one facility – a hallmark – which serves as a hub for maintaining the presence and activity of the residents. Similar analyses were described by Waldemar Siemiński in the 1930s. A city filled with a human presence in the streets is safe, interesting and made use of, but what if the population of cities is becoming too large?

The United Nations indicated only one city located in a developing country – Buenos Aires – in a list of eight of the world's largest metropolises in 1950 (Pawlak & Pawlak, 2010). A UN Department of Economic and Social Affairs publication in 2019 (*The 2018 Revision of World Population Prospects*; UN, 2019) identified the 15 largest cities in the world based on population studies (Table 3.1).

Urbanisation has accelerated significantly since the 1960s. Initially, it spread across only those countries of the Latin America region that adopted an economic development policy based on the principles of import substitution and the promotion of industrial locations. An example is Mexico with its characteristically rapid population growth rate in the capital (Czerny, 2014). The country has a high percentage of urban dwellers. In 2019, Mexico's population amounted to 132,328,000 (MSZ, 2018). The average age has increased from 22.4 years in 1960 to 31.4

Table 3.1 Ranking of the 15 largest world cities by population (in 2018)

Ranking	City	Population (in millions)
1	Tokyo	37.5
2	Delhi	28.5
3	Shanghai	25.6
4	São Paulo	21.7
5	Mexico City	21.6
6	Mumbai	20.8
7	Cairo	20.1
8	Lagos	20.0
9	Dhaka	19.7
10	Beijing	19.6
11	Osaka	19.3
12	Karachi	16.1
13	Istanbul	14.1
14	Guangzhou	12.9
15	Kinshasa	12.9

Source: Department of Economic and Social Affairs (2019).

Table 3.2 The population of Mexico: Selected years

Year	Average age	Birth rate	Population density (people per km²)	Urban population (and the percentage of town dwellers in relation to the total population)
1960	22.4	6.77	20	19,374,507 (50.8%)
1980	22.4	4.84	36	46,013,308 (66.3%)
2000	26.3	2.72	52	76,006,974 (74.7%)
2016	30.5	2.19	66	101,416,318 (79.5%)
2017	30.8	2.16	66	103,052,000 (79.8%)
2018	31.1	2.13	67	104,666,000 (80.3%)
2019	31.4	2.10	68	106,263,000 (80.3%)
2020	31.7	2.07	69	107,840,000 (80.6%)

Source: INEGI (2019).

today. The urban population increased from 76,006,974 (74.7%) in 2000 to 107,840,000 (80.6%) in 2020. Mexico's population density changed from 35.7 in 1980 to 67.3 in 2018 (INEGI, 2019) (see Table 3.2).

3 Methods and Data

Urbanisation is a consequence of globalisation. Globalisation refers to trends in the world economy, politics, demography, social life and culture, with the spread of similar phenomena, regardless of the geographical context and the degree of economic advancement of a given region. It makes no sense to criticise the process of creating new cities because globalisation will make it impossible to stop. Instead, analysis of the available data allows us to systematise the problems in cities and try to solve them in the spirit of the concept of sustainable development.

A triangulation method was used in this study. This is a method often used in social research to ensure a higher quality of research and a reduction of measurement error. It collects data using two or more methods (e.g. observation together with content analysis) and compares and combines the results, also combining qualitative and quantitative methods. In this article, data triangulation is used in a special way – presenting and comparing the analyses and research carried out on different populations, at different time intervals and in different places.

Globally, the urbanisation index increased from 29.8% in 1950 to 37.9% in 1975 and 47.2% in 2000. In 2007, the urban population exceeded the number of rural residents worldwide. In 2010 the urbanisation index reached 57.2%. According to forecasts, by 2025 the proportion of the urban population in developed countries will not have increased much (up to 78%), while in the Global South the increase will be enormous. It is estimated that between the years 2000 and 2030 the population in developing countries will increase by 2 billion (Czerny, 2014), after which the population will start to decrease. According to forecasts in the report by the United Nations Department of Economic and Social Affairs, by 2050 there will be almost 10 billion (9.7 billion) people in the world (Jędrzejewska, 2017). This increase will be a consequence of population growth in Africa and Asia. In Africa, the population increases by 2.5% each year. Australia and Oceania rank second in terms of population growth with a rate of 1.4%. The population growth of South America and Asia is at a similar level. North America is less dynamic (0.7%), while Europe is the weakest (annual growth of 0.1%). Both of these continents, although rich, have the highest percentage of elderly people (UN, 2019). In Europe, 19% of the population is 65 or older and the figure is 16% in North America (UN, 2019). According to forecasts, this percentage will rise to 28% in 2050 on both continents (UN, 2019).

Many factors determine the pace and size of the spatial expansion of cities. The greatest contrasts are visible in Africa and Asia. Burundi, Malawi and Uganda are the countries with the lowest degree of urbanisation in the world, while Libya and Gabon are among the most urbanised. In Singapore, Qatar and Kuwait, 100%, 99% and 98% of the population live in the city, respectively (MEN, 2019), while Sri Lanka, Nepal and Cambodia (MEN, 2019) are the least urbanised (16%, 18% and 21%, respectively). There is a relationship between the degree of urbanisation and the degree of development. In developing countries, cities are experiencing population growth and people are migrating intensively. Cities provide opportunities for easier access to food, to the labour market and to education and health services.

The expansion of urban agglomerations often occurs in a way that is unreasonable and harmful to societies. The spontaneous and uncontrolled urbanisation of countries in Asia and Africa is creating slums in cities that are unable to accommodate and meet the needs of their populations. A large proportion of people settling in cities in the developing countries have little or no income, so they choose unattractive parts of the city to live in.

Unattractive areas are often dangerous (floodplains, steep hillsides, for example). Over time, they grow to the size of housing estates and districts. Slums are areas on the outskirts of cities as well as near the centre, in the vicinity of the richest districts. In Rio, this is how every fifth person lives (Xavier & Magalhães, 2003). Slums in India (called *bustees*) and in Latin America (called *favelas*) are districts of misery. In cities such as Lagos, São Paulo and Calcutta there is no population policy. Poor economic conditions together with fast population growth increase the urban crises of Jakarta, Dhaka and Karachi (Dzieciuchowicz, 2009). Cairo is a city with moderately developed urbanisation processes. Attempts to slow down the pace of demographic growth are an element of the governmental policy and show the awareness of the problem and its consequences.

The United Nations Human Settlement Programme (UN-HABITAT), focused on the material aspect, created a definition of slum households. It states that a slum is a group of people living in an urban area under the same roof who do not have at least one of the following: a permanent dwelling, adequate living space (no more than three people per room), easy access to drinking water, adequate sewage infrastructure or ownership of the premises (no forced evictions) (Barber, 2014). One in three inhabitants of cities in the world lives in a slum. In Africa the proportion is much higher – six in 10 inhabitants (UNICEF, 2012). In China, 37.7% of the urban population live in slums, in India, 55.5% and in Brazil, 36.6%. In Pakistan the figure is 73.6%, in Nigeria, 79.2%, in Bangladesh, 84.7%, in Tanzania, 92.1% and in Ethiopia, 99.4% (Barber, 2014). Sub-Saharan Africa is characterised by the presence of the worst slums in the world. The Third World countries are the second enclave of slums. The reason for such a large number of slums is the rapid and uncontrolled growth of large cities. The inequalities causing the creation of slums and districts of poverty are not only a sad reality but also a cause of differences between people in terms of their talents, ambitions and goals – they are the traits of urban life that deeply distort them (Barber, 2014).

4 Results and Discussion

Regardless of the presence of slums or exclusive neighbourhoods, city authorities must deal with similar problems related to high population density: the development of infrastructure, the provision of housing, access to water, public transport, energy, security and waste disposal. However, the consequences of urbanisation are not a product of the present. Even in ancient Rome, under Julius Caesar's rule, the Senate issued a regulation specifying the appropriate time of day for travelling by certain means of transport. The regulation was introduced to reduce noise. In 16th-century London, Queen Elizabeth I issued a decree in which she banned burning hard coal in fireplaces due to high air pollution in the city. There was a death sentence for disobeying the Queen's order (Mirowski, 2016). These examples provide insight into some of the measures that were taken to control the negative impacts of urbanisation. Both urbanisation and demographic phenomena can be transformed into socially and economically beneficial processes, conditioned by rational policies of the central and regional administration. There is still a shortage of political instruments and economic conceptions taking into account long-term goals which would solve the contradictions of globalisation and reduce the disproportions of development in the world. In recent years, the problem of the intensity of urban land use has become a leading subject of urban debates. Rational intensification of the use of urban areas is currently the only solution to the problems resulting from the development of the world's cities.

Until the 19th century, European cities developed in a concentrated way. External layers grew successively around the historical market square – the centre. Residents accumulated on the smallest territory possible. The reasons for the compactness of cities were security and social hierarchy. In the era of globalisation and modern urbanisation processes, the compact city model envisages stopping the process of dispersion of buildings and implies the revitalisation of degraded areas and the reduction of car transport in favour of public transport. The compact city model has been included in the European Spatial Development Perspective. The Athens Charter (Europejska Rada Urbanistów, 2003) introduced a new value – the so-called functional city. The counter-urbanisation of cities and the urbanisation of villages are the main assumptions of the functional vision of a territorial unit.

A return to the idea of a compact city took place in the 1960s. The refreshed concept of the compact city became an instrument of sustainable development. The intensity of the development of the compact city saves land. The careful management of public spaces – much smaller than in scattered cities – makes it possible to maintain their quality. A compact city favours effective satisfaction of residents' needs, which is reflected, among other things, in the ease of access to services due to short distances and the independence of residents from car transport. It introduces ecology to the city. A compact city has great social significance. It is an attractive space for residents and has an integrated function. The reduction of travel distances and spatial closeness of the inhabitants has a positive impact on the relations and bonds of urban communities. However, some of the proclaimed advantages of a compact city are questionable. Urban planning processes are also often the result of local, unique factors. In Delhi, the highest intensity of development occurs in the outskirts of the city (four times higher than in the central districts). Urban planners call it the inverted compact city. Ashok Kumar claims that this is a consequence of political settlement decisions made in the 19th century on the spatial layout of the city, whose centrally developed area was inhabited by colonisers only. The 20th century saw the start of vibrant and great migrations. In the 1980s alone, the population of Delhi increased by 30%. A large proportion of the new residents (more than 1,300,000) (Kowalewski, 2005) intensively populated the surrounding central districts, which completely changed the city space. As a consequence, the policy of intensification of development – instead of creating a compact city – led to the creation of slum districts. Barrios in Caracas – a district of poverty that disintegrates the compact city space – is another example of the consequences of misguided urban policy. Caracas has the highest murder rates in the world and although the socioeconomic indicators have improved in the city compared to those in 2010, this is a consequence of informal criminal links rather than rational development policies of the city and its parts. The assumptions of the concept of sustainable development largely coincide with the principles of modern urban planning. They include the postulates contained in the Athens Charter, the well-known Habitat Agenda or the declarations of the Urban 21 Congress. The discussion on the compact city model shows how complex the problem is. An intensively developed, environmentally friendly city, which is also comfortable and friendly for its inhabitants, requires the application of various solutions, which must be consistent with the time, place and manner of their application.

5 Conclusion

The city as a demographic and economic fact (Barber, 2014) is an area of compromise and complementary solutions. These compromises must take into account the changing of habits established for generations. In Denmark, 200 inhabitants are sufficient for a region to become a city, while in India there must be at least 20,000 inhabitants. In 1950 there were only two megacities (units of over 10 million inhabitants) in the world – New York and Tokyo. In 1995 the number rose to 14, to 33 in 2018 and in 2030 there are expected to be over 43. The explosion of cities that has been going on for decades cannot and should not be stopped. It is an implication of justified and necessary changes in the era of globalisation. The changes should include production technologies, which will affect the global economy. It is impossible to reconcile the contradictions between urban development and consumption and economic objectives. In urban planning, there are many areas where improvement can be achieved in a logical and evolutionary way. However, this requires rational actions. Strategic assumptions included in development plans should programme urban development in the perspective of the future, not focus on current issues, as is currently the case. While it is important to remove the effects of uncontrolled urbanisation, it is much more important to prevent them.

References

Barber, B.R. (2014) *Gdyby burmistrzowie rządzili światem. Dysfunkcyjne kraje, rozkwitające miasta*. Warsaw: MUZA S.A.

Czerny, M. (2014) *Stare i nowe w przestrzeni miast Ameryki łacińskiej. Aktorzy i kontestatorzy zmian* (*Old and New in the Space of Latin American Cities. Actors and Contestants of Change*). Warsaw: Wydawnictwo UW.

Department of Economic and Social Affairs (2019) *World Urbanisation Prospects: The 2018 Revision*. New York: United Nations.

Dzieciuchowicz, J. (2009) Obraz demograficzny świata u progu XXI wieku (The demographic picture of the world at the dawn of the 21st century). *Acta Universitatis Lodziensis*. September.

Europejska Rada Urbanistów (2003) *Nowa Karta Ateńska 2003. Wizja miast XXI wieku* (*La Nouvelle Charte d'Athenes 2003/The New Athens Charter*). Florence: Alinea.

Hodder, I. and Marciniak, A. (2015) *Assembling Çatalhöyük*. Wakefield: Charlesworth Press.
INEGI (2019) *Población*. See https://www.inegi.org.mx/ (accessed 10 April 2020).
Jacobs, J. (1961) *The Death and Life of Great American Cities*. New York: Random House.
Jędrzejewska, W. (2017) Nowy raport ONZ zakłada, że populacja ludzka osiągnie 9.7 miliarda osób do roku 2050 (A new UN report assumes that the human population will reach 9.7 billion by 2050). *WP Tech*, 23 June. See https://tech.wp.pl/nowy-raport-onz-zaklada-ze-populacja-ludzka-osiagnie-97–miliarda-osob-do-roku-2050–613670 0030826625a (accessed 2 January 2020).
Kolińska, X. (2009) Qaramel najstarsze miasto świata. Skarby sprzed 13 tysiącleci (Qaramel – the oldest city in the world. Treasures from 13 millennia ago). *National Geographic Polska*, 16 September. See https://www.national-geographic.pl/historia/ qaramel-najstarsze-miasto-swiata (accessed 1 January 2020).
Kowalewski, A.T. (2005) Rozwój zrównoważony w procesach urbanizacji (Sustainable development in urbanization processes). *Nauka* 1. Warsaw: PAN.
MEN (2019) *Urbanizacja i światowe metropolie* (*Urbanization and World Metropolises*). See https://epodreczniki.pl/a/urbanizacja-i-swiatowe-metropolie/D7nJ8MsAp (accessed 30 December 2019).
Mirowski, A. (2016) *Podręcznik dobrych praktyk w zakresie doboru i wykorzystania odnawialnych źródeł energii* (*Manual of Good Practices in the Selection and Use of Renewable Energy Sources*). Warsaw: ARL Mirowski.
MSZ (2018) *Profile krajów* (country profiles). *Explore All Countries. Mexico. North America* (2021). See https://www.cia.gov/the-world-factbook/countries/mexico/ (accessed 9 July 2021).
Pawlak, N. and Pawlak, J. (2010) *Zrównoważony rozwój miast* (*Sustainable Urban Development*). See http://www.agenda21.waw.pl/index.php?option=com_content& view=article&id=55&Itemid=53&showall=1 (accessed 2 February 2020).
UN (2019) *The 2019 Revision of World Population Prospects 2019*. See https://population.un.org/wpp/ (accessed 20 March 2020).
UNICEF (2012) *Podsumowanie Raportu UNICEF. The State of the World's Children 2012: Children in an Urban World*. See https://www.unicef.pl/O-nas/Aktualnosci/images/stories/aktualnosci/foto/raport_state/2012_02_28_podsumowanie_raportu_unicef_sowc_2012.pdf (accessed 3 January 2020).
Wirth, L. (1996) Urbanism as a way of life. *American Journal of Sociology* 44 (1), 1–24.
Xavier, H.N. and Magalhães, F. (2003) The case of Rio de Janeiro. In *UN-Habitat. Global Report on Human Settlements 2003: The Challenge of Slums* (pp. 195–228). London: Earthscan.

4 Nutrition Insecurity in a Developed World: Challenges for Health Promotion and Further Directions for Sustainable Development

Urszula Śliwka

The aim of this literature review is to identify gaps in existing knowledge on whether poverty and the insecurity of good quality food, which shape health outcomes in the developed world, could be sustainably improved for communities and individuals as preventative healthcare. While billions suffer nutritional deficiencies because of the lack of available food, significant numbers worldwide are overweight or obese. Overconsumption of calories can occur alongside poor nutrient intake and malnutrition regarding essential micronutrients. Research has shown that this is often associated with economic disproportions in developed countries as well as low-quality food production and agriculture. The Sustainable Development Goals from the Global Nutrition Report focus on the major areas of nutritional concern, which are linked to preventable and poverty-related health challenges such as heart disease and diabetes, undernutrition and obesity, iron deficiency and low birthweight. This literature review shows that governments' food and nutrition policies, while undertaking action to achieve sustainable development, can simultaneously stress the need for a more individualised approach in identifying household food insecurity and socioeconomic disadvantages for preventative healthcare for all. In order to improve well-being, sustainable development must overcome several limitations directly linked to the dietary habits, lifestyle choices and mental health of individuals as well as the local ecosystem and agriculture. Further development and innovation systems that link research and local knowledge regarding nutritional insecurity are needed to improve health promotion for individuals and the whole community.

1 Introduction

Over the years, inappropriate nutrition has been associated with an increase in life-threatening medical conditions such as obesity, coronary disorders, diabetes and cancer. For the human body to function well, there should be an adequate supply of both macronutrients and micronutrients (Crume *et al.*, 2016). Macronutrients help fuel the body by providing calories, whereas micronutrients (vitamins, minerals and other necessary chemicals) aid physiological processes (Vernarelli *et al.*, 2018). Food intake in most high-income nations involves an exaggerated consumption of carbohydrates and fats but a limited intake of micronutrients (Mozaffarian *et al.*, 2018). Carbohydrates and fats are cheap and are prepared so that they are sweet, promoting overconsumption, especially in people of low income (Vernarelli *et al.*, 2018). It is a matter of concern that while there is a proliferation of obesity, many individuals throughout the world are undernourished.

People can experience different levels of food insecurity. It can be at a moderate level, which is connected with the low quality and/or quantity of food consumed due to, for example, a lower socioeconomic position, or the most extreme and life-threatening hunger caused by poverty. Nutrition insecurity and micronutrient malnutrition are also present in the developed parts of the world and are directly linked to development and economic growth. This modern type of nutrition insecurity is either an effect of the massive amount of available processed food, which has little nutritional value and a high number of calories, or is caused by conventional agriculture, which utilises chemical fertilisers to enhance plant growth and does not focus on delivering nutrient-dense foods.

Determining whether poverty and food insecurity are fundamental elements of sustainable development goals could help us better understand how economic growth shapes the global and local food system and how it could prioritise the health of communities and individuals from a health preventative perspective.

Topics like poor human nutrition, obesity and malnutrition, barriers to eating a nutritious diet and agricultural intensification, as well as poverty and food insecurity, are related to each other, and all of them shape health outcomes. Studying the background of and mechanisms between rapid economic growth, the development of obesity and the high level of poor nutritional status among people in developed countries is an important matter, and in both the southern and northern hemispheres has the potential to improve food insecurity while reducing incidences of mortality and morbidity linked to poor nutrition.

2 Background

In many countries, life expectancy is increasing, and this has been happening for the last 200 years (Chetty *et al.*, 2016). In the early years the increase was realised mainly by putting mechanisms in place that decreased childhood mortality, but today a longer lifespan is achieved by altering factors that affect life in later years (You *et al.*, 2015). Statistics indicate that by 2040, 25% of people will be over 65 years (Colby & Ortman, 2015). However, this advancement in age is characterised by an increase in debilitating health conditions like diabetes, heart (WHO 2014). There is a remarkable increase in the effects of illnesses over the last 10 years on individuals' lives (You *et al.*, 2015). Many non-communicable diseases (NCDs) could be prevented by the intake of the correct nutrients. However, access to a proper diet has been a perennial challenge globally (Alsaffar, 2016).

In their report, the Food and Agriculture Organisation of the United Nations (FAO, 2019) showed that poverty reduction is not always occurring in some developed countries and that fast economic growth and development can be harmful to the food system in term of food security and nutrition.

Many nations, especially high-income ones, produce enough food for their citizens. However, the degree of micronutrients in these foods is below the recommended threshold. Such diets fail to appropriately support bodily metabolic and physiological processes (Beal *et al.*, 2017). Research has indicated that three-quarters of adults ingest an insufficient quantity of vitamins, while half of children under five years do not meet the recommended vitamin intake (Yakoob *et al.*, 2016). As a consequence of incorrect diet, obesity cases are on the increase, even in developing countries, and are becoming increasingly prevalent in children.

Maternal malnutrition is also of concern as it leads to micronutrient deficiencies, stunting, underweight births and vitamin A and zinc deficiencies. It leads to over 850,000 neonatal deaths and over 3 million child deaths annually (Yakoob *et al.*, 2016). If no interventions are undertaken, maternal malnutrition affects both the mother and the child for many years (Bhutta *et al.*, 2013). Improving nutrition is critical for the promotion of health. Focusing on adolescent health is crucial as it is the foundation for achieving good health for women, as well as for children who will soon reach child-bearing age.

The quantity of food being produced worldwide is sufficient for each person globally, but there are various hindrances such as low economic status that discourage individuals from consuming nutritious food (Johnston *et al.*, 2014). Indigenous fruits and vegetables are necessary for enhancing health in the rural population and promoting poverty eradication by raising the growers' earnings as well as voiding the food security challenge. However, these foods are not yet embraced globally (Ashraf *et al.*, 2018).

If the condition of healthy eating is to be realised, then agricultural objectives must be restructured such that a particular focus is made of diversity, at the same time emphasising sustainability (Jones & Ejeta,

2016). Agricultural intensification is critical for addressing these challenges (Altieri *et al.*, 2012). Raising food production output by 1% is likely to lower the poverty index by 2% (Oseni *et al.*, 2014). However, agricultural intensification creates some problems. The overuse of agrochemicals and fertilisers leads to pollution of the environment (Carvalho, 2017). Andersen *et al.* (2015) suggest that turning to traditional approaches of farming could be better for reducing pollutants compared with conventional methods, despite having a lower yield. This could be additionally supported by both ethical and practical rethinking when food insecurity, in term of limited access to foods high in nutrients and free from chemicals, matters and affects people's health status (Śliwka & Przybyłowicz, 2017).

3 Human Nutrition

3.1 A negative effect of an inadequate supply of micronutrients

When humans consume an inadequate amount of micronutrients (less than the recommended dietary allowance, RDA), normal metabolic processes may not take place. The body needs all the micronutrients in proportionate quantities in order to function well, and when one constituent is missing or inadequate metabolic function is disrupted, this may cause serious health issues including degenerative diseases (Ladeira *et al.*, 2017). The ageing process is highly dependent on the state of the mitochondria. Lack of micronutrients in the body promotes mitochondrial decay, resulting in degenerative conditions such as cancer (Ladeira *et al.*, 2017). Studies have indicated that restricting the consumption of macronutrients while supplying optimal levels of micronutrients promotes the life expectancy of various animals (Mattison *et al.*, 2017). This perspective of addressing deficiencies in micronutrients, with a focus on the fortification of food, coupled with the desire for better health, might have positive outcomes in improving health.

3.2 Magnesium deficiency

In many countries, most individuals take in less than the recommended quantities of magnesium. For example, in the United States, 55% of adults do not ingest an adequate quantity of magnesium, being short of the estimated average requirement (EAR) expectations (Quann *et al.*, 2015) (see Table 4.1). Consumption lower than EAR is more pronounced in households with a low income and in some demographic groups. The prevalence of magnesium deficiency is 78% in male teens, 90% in female teens, 82% in obese African Americans and 80% in the elderly (Waern *et al.*, 2015). Barbagallo and Dominguez (2015) contend that the insurgence of childhood obesity, when coupled with an inappropriate consumption of macronutrients, heightens the risk of type 2 diabetes.

Table 4.1 Selected micronutrient inadequacy in the United States

Micronutrient	Affected group	% Undernutrition
Minerals		
Iron	Women aged 15–50 years	16
Magnesium	Everyone	55
Zinc	Everyone	15
Vitamins		
Vitamins B6	Women over 70 years	48
Folate	Adult women	18
Vitamin E	Everyone	92
Vitamin C	Everyone	33

Source: Quann *et al.* (2015).

Micronutrients also play a critical role in the body as they are associated with insulin action. In adults, an inadequate supply of magnesium in the cells is correlated with insulin resistance, reduced insulin secretion and glucose intolerance (Singh *et al.*, 2015). Surveys in adults show that a reduced consumption of magnesium and reduced serum magnesium may be critical in the onset of type 2 diabetes. Magnesium also plays a crucial role in children. Obese children tend to possess reduced quantities of serum magnesium compared to lean children (Vernarelli *et al.*, 2018).

3.3 Iron deficiency

Obesity and iron deficiency are two epidemics that affect millions of people worldwide. Their high prevalence is not by chance but has been found to be scientifically associated in the manner in which they influence each other (Zhao *et al.*, 2015). Aigner *et al.* (2014) point out that the complex processes in energy homeostasis in adipose tissue and the liver are associated with transformations of iron homeostasis. There is an inverse correlation in the iron status and excessive overweight, whereby substantially reduced serum iron concentration is observed in obese individuals as opposed to their normal counterparts.

3.4 Zinc and other trace elements

Trace elements play a part in preventing pro-inflammatory processes and peroxidation which are critical processes in the onset of obesity, insulin resistance or hypertension. Obese individuals have decreased serum zinc concentrations compared to individuals with normal weight. These results are consistent in all age groups, including children under 16 years (La Frano *et al.*, 2014).

Azab *et al.* (2014) established that serum zinc and selenium levels were substantially reduced in obese children; on the other hand, serum copper, cholesterol and triglycerides were higher in obese children compared to their normal counterparts. An inverse correlation was also observed between serum leptin and zinc concentrations in obese participants. Zinc is especially crucial in the metabolism of hormones that affect the mechanism of the development of obesity (Aigner *et al.*, 2014). The level of zinc in the body is associated with serum leptin levels which take part in the regulation of satiety in which obese individuals have higher serum concentrations and consequently decreased satiety. Copper is a critical element of antioxidant enzymes that manifest in the body to clear free radicals which promote the onset of heart diseases. Therefore, when copper is not balanced in the blood, its metabolism may invoke the onset of hypercholesterolemia (Azab *et al.*, 2014).

4 Current Nutritional Condition

Many developed countries have a sufficient supply of food. However, there is a significant inadequacy of micronutrients in their staple food (Beal *et al.*, 2017). Although many people access enough foodstuffs, such a diet may not necessarily be described as healthy (Yakoob *et al.*, 2016). Diet surveys show that 75% of adults do not consume the recommended quantity of vitamins (Leblanc *et al.*, 2016); moreover, about 50% of children under five years consume a small amount of vitamins (Yakoob *et al.*, 2016). An examination of food intake trends in children revealed that between one in two and seven in ten children aged 9–19 years fail to consume the recommended threshold of vitamin A, riboflavin, vitamin B12, thiamine, vitamin C, pantothenic acid and niacin (Beal *et al.*, 2017).

4.1 Obesity and undernutrition

Between 1975 and 2016, cases of childhood obesity increased threefold. By 2016, there were almost 2 million overweight adults worldwide, of whom 650 million were obese (WHO, 2019). In the same period, 40% of overweight people and 13% of all adults were obese. In the past decades, childhood obesity was associated with affluence and therefore was anticipated in high-income countries (WHO, 2019). However, there has been a notable upsurge in obesity frequency in nations with average or low-income status. Even in Africa, where most countries are developing, obesity rates have doubled since 2000 (WHO, 2019).

4.2 Barriers to eating more of a healthy wholefood diet and less processed foods

According to the FAO (2019), development and economic growth are not always directly linked to the reduction of poverty for too many countries that are still experiencing food security and poor nutrition. Nguyen *et al.* (2019) compare the food industry to the tobacco industry by alluding to the fact that in the 1950s the established tobacco companies conspired to mislead consumers regarding the potential dangers of smoking. The companies heavily invested in sensitisation through the media to mislead the public. They also had access to elite lawyers and influenced some key government officials. All these strategies were aimed at convincing people that the alleged effects of tobacco smoking were not so severe and they even went ahead with repackaging the same products and labelling them as filtered or low-tar cigarettes. The efforts were geared towards maintaining high profits while disregarding the health implications. This scenario has been duplicated in the food industry. However, there is a difference in that food is necessary for everyone, whereas one can choose not to smoke.

The food industry consists of the companies that manufacture, process, distribute and serve food materials and beverages, including food supplements. These companies greatly influence government food policies and do so for the sole purpose of preserving their business as opposed to promoting healthy eating (Nestle, 2013).To promote their business, food sellers convince consumers as to why they should use one food product over another. They do this in a manner that is complex, taking advantage of the fact that individuals choose what they eat based on economic, social and cultural contexts (Nguyen *et al.*, 2019). Food companies have mastered several strategies to sell their products. For instance, they focus on improving the taste of the food. Foods are made visually attractive, aromatic and presented in a wide variety of ways (Chattopadhyay *et al.*, 2014). Companies claim to add value by processing the foods. The farmers often receive a small fraction of the amount the individual pays for the packaged products. For instance, potatoes from the farm are cheap and healthy but are processed into expensive and unhealthy potato chips (Nguyen *et al.*, 2019). The companies have taken advantage of the changing world where people are getting busier, and therefore present them with value-added fast foods for their convenience. Mealtimes often find people outside their homes, thus forced to consume readily available foods (Nestle, 2013).

Another strategy that companies employ is keeping consumers puzzled. Press reporters often present their information in a biased manner depending on the motive and other vested interests, and thus they can fail to maintain a clear and focused stance. The resulting confusion is geared towards benefiting the food processors and the media at the expense of the health of the consumers (Nestle, 2014).

The ultimate approach that companies employ is to emphasise eating more by consistent advertising. They are competing to convince consumers to buy more of their foodstuffs and are always expanding their customer base, including children and other minority groups. They undertake fierce advertising as well as formulating products designed for specific target groups (Nestle, 2014). A vast amount of money is pumped into adverting. In the United States, food companies pump over ten billion dollars annually into media advertisements. The advertised foods are often highly processed, and customers are enticed by good promotions (Folkvord *et al.*, 2016). Despite the high marketing expenditure, the sales far exceed the expense.

5 Improving Nutrition

5.1 Discouraging processed food, embracing wholefoods

Sales promotion and advertising impacts the minds of the consumers strongly and results in repeat purchasing. Although free samples influence customers, buy-one-get-one-free dramatically affects customer purchasing trends (Oladepo & Abimbola, 2015). Such persistent adverts induce individuals to consume more. A good percentage of individuals are affected by addictions, not only to addictive substances but also to food. Some adults and children consume more food than necessary, and in the long run develop health problems. Cases of emotional eating whereby individuals are overwhelmed with intense cravings for food and helplessness in controlling the quantity of high-calorie food consumption, ignoring their known adverse health impacts, are substantial (Gearhardt *et al.*, 2011). The ultimate effect is overweight or obesity, which is a burden not only to the individual but also to the government (Piernas *et al.*, 2016). Increased incidences of obesity have been observed in individuals with the highest poverty index and the lowest educational levels. Energy density correlates to energy cost, as high-calorie foods consisting of sugars, refined cereals and fats are most affordable for consumers (Schlüssel *et al.*, 2013).

Clemens *et al.* (2012) propose that the players in food production and distribution should take into consideration consumers' food intake behaviours and help them consume whole grains with an adequate fibre content. To achieve this requires dividing consumers into smaller groups and providing them with a realistic approach to overcoming barriers to healthy eating. It could be beneficial to address this problem early by providing healthy wholefoods to children, thus preventing the future development of diseases. Clemens *et al.* (2012) further assert that parents and caretakers should not be manipulated by the child's preference for processed and often over-sweetened foods at the expense of whole grains. Presenting a diversity of high-fibre foods early in life would make them more appealing. Another option is to include high-fibre ingredients in the foods commonly consumed by children. For example, one could use whole wheat flour in the baking of pizza and crackers.

Bodies in charge of dietary guidelines should ensure that manufacturers correctly label the fibre content on their packaging. Lack of proper labelling or giving false information leaves consumers confused, leading to unintended consumption of food with a low fibre content. Howlett *et al.* (2010) claim that over 50% of the grains consumed are acquired from whole plant grain and the rest from the enriched grains. Overall, grains contain significant quantities of micronutrients that provide enhanced health benefits to consumers. Consumers should, therefore, monitor not only their calorie intake but also the fibre and micronutrient consumption.

5.2 Improving maternal and child nutrition

Maternal undernutrition leads to micronutrient deficiencies, stunting, underweight births and vitamin A and zinc deficiencies. It is a source of concern as it accounts for over 850,000 neonatal deaths and over 3 million child deaths annually. The number of stunting cases has been reducing over the years, but the issue is still significant, as 170 million foetuses were found to be stunted in 2011 (Black *et al.*, 2013) and 151 million in 2015 (WHO, 2017). Maternal nutrition and child obesity are also on the rise, particularly in developing nations (Lim *et al.*, 2012). Maternal malnutrition has a trickle effect in that it affects the individual at various stages of development. Therefore careful dietary consideration is necessary at critical stages of life. Particular emphasis should be placed on enhancing diet in neonates, infants, children under five years of age, school-age children and adolescents (Bhutta *et al.*, 2013).

5.3 Focusing on adolescents' health and women of reproductive age

Addressing adolescent health is crucial as it is a precursor to attaining balanced health for women and children given that a significant number of girls under the age of 18 years get pregnant or are married. Opportunities may also be presented to address the under-consumption of micronutrients and the control of obesity. Gestational folic acid supplementation is critical in improving average birthweight and the reduction of blood-related challenges. Iron supplementation during pregnancy has shown a 70% reduction in anaemia (Lim *et al.*, 2012). Maternal calcium supplementation has indicated a potential for lowering gestational hypertensive disorders significantly. Consuming iodised table salt could address issues resulting from iodine deficiency (Bhutta *et al.*, 2013).

6 Embracing Agriculture to Produce Nutrient-dense Foods

The new Sustainable Development Goals underscore the importance of countries' endeavours to eliminate food insecurity, to provide adequate nutrition to all and to embrace sustainable agriculture. According to Johnston *et al.* (2014), the quantity of food being produced worldwide is sufficient for each person globally, but some individuals still have to overcome various obstacles in order to consume the right diet. Such barriers include cultural beliefs, low economic status and personal preferences. Many nations have made positive steps towards addressing malnutrition, but most of them, particularly in the developed countries, contend with increasing cases of obesity and other NCDs (Joy *et al.*, 2017). The issue of addressing both undernutrition and obesity has had a retarding effect on economic progression, the restructuring of food production and consumption, and the improvement of health services (Alsaffar, 2016).

The complex nature of the correlation between the consumption of food and health and the focus on the effect of some food compounds on health have posed a challenge to the optimisation of food production. Nonetheless, many scholars and health experts have continued to emphasise that dietary recommendations should be made based on foodstuffs rather than the natural nutrients (Jones & Ejeta, 2016). Taking the food-based approach, scientific studies have indicated that the consumption of simple food materials, in particular a variety of cereals and vegetables, has a positive outcome on health (Katz & Meller, 2014). Agriculture would be sufficient to meet this simple guideline. Unfortunately, many nations have not realigned their agricultural systems to focus on diversity rather than the quantity of certain preferred foods. Throughout the world, three main crops, maize, wheat and rice, constitute over 70% of food consumption. This creates a deficit for the production of vegetables, fruits and pulses (Siegel *et al.*, 2014). Moreover, there has been an insurgence of mass production of animal feeds, industrial manufacture of ingredients such as sugar sweeteners, and biofuels, all of which is done at the expense of the production of healthy foods (Katz & Meller, 2014).

If the desire to consume healthy food is to be realised, then agricultural objectives must be restructured so there is a particular focus on diversity, and at the same time an emphasis on sustainability. New systems in agrarian research are necessary to enhance affordable methods of production of vegetables, fruits and pulses. At the same time, mechanisms have to be formulated to regulate markets and create food value logistics to enhance production capabilities. These efforts are crucial for promoting the economic power of farmers as well as ensuring an adequate supply of vegetables, fruits and pulses, while preserving food components in these products once they are harvested (Jones & Ejeta, 2016). These undertakings can only be feasible if concurrent mechanisms are put in place in the food chain to bring about demand for nutrient-dense foods.

6.1 Use of fertilisers and pesticides in crop production

All nations are now focused on reducing poverty and hunger levels. While the amount of land for cultivation remains relatively constant, the need for food is increasing. Agricultural intensification whereby more food crops are grown on the land under cultivation would go a long way towards addressing these challenges (Altieri *et al.*, 2012). Oseni *et al.* (2014) argue that increasing food production by 1% would result in a decrease in poverty by 2%. However, maximising agricultural output presents some challenges. Consumers are gaining increased insight into the ecological effects of conventional agriculture, which results in the contamination of the end-product. In order for crops to be unpolluted, the soil, water and air should be free of pollutants. These contaminants include pesticides, heavy metals and components of fertilisers, antibiotics and radioactive isotopes. The unregulated application of agrochemicals and some agricultural practices leads to contamination of the water and soil and significantly impacts the environment and ultimately health. The pollutants are of small quantities, and therefore their effect on human health occurs after many years. Agrochemicals that typically find their way into water sources have been linked to the proliferation of chronic kidney disease (Wimalawansa & Wimalawansa, 2014). Modern agriculture is heavily dependent on agrochemicals, driven by the desire to curb hunger, despite recognition of the potential dangers of such chemicals to the body. Often fertilisers, pesticides and herbicides can be acquired from retail outlets without any control over the quantity one can safely apply (Carvalho, 2017). The unsustainability of the availability of organic food is also an ethical matter as the majority of the population across the globe are not able to grow their own food, and organic food is too expensive for them to buy and feed to their families (Śliwka & Przybyłowicz, 2017).

Andersen *et al.* (2015) suggest that turning to traditional farming approaches could be beneficial, as the resulting pollutants had superior properties and yields compared with conventional methods in which intensive practices are undertaken. However, the overall return is often higher in conventional agriculture than in organic farming. Seufert *et al.* (2012) suggest the rewilding of crops to revitalise genetic diversity.

6.2 Making use of underutilised vegetables complementary to micronutrients and addressing malnutrition

As the population increases and demand for food soars, underutilised crops, particularly fruits and vegetables, become necessary to fill the food gap. These crops are essential for addressing malnutrition and enhancing health in the rural population and promote poverty eradication by raising the growers' earnings as well as voiding the food security challenge (Ashraf *et al.*, 2018). Many of these crops are found in various parts of the world, but they are often grown locally, serving as food, local medicine and fodder for the local population. These indigenous plants have more food nutrients than widely consumed vegetables like cabbage and tomatoes. Besides having important micronutrients, they are largely disease and pest resistant, making their cultivation easier, especially for the low-income population (Nandal & Bhardwaj, 2014).

Lentils and kale are two crops which, when used in combination with other foods, can bring about positive outcomes in tackling obesity and micronutrient malnutrition. Lentils contain a substantial amount of proteins (up to 30%) and about 5% of minerals as well as vitamins and prebiotic carbohydrates (Migliozzi *et al.*, 2015). Lentils have many health benefits, and since they help in nitrogen fixation and can be grown in quantity, they can appropriately be used in micronutrient biofortification. However, in the past, practitioners preferred other crops to lentils. Even though studies have been undertaken regarding commercial production and use of lentils and other pulses, still the production and distribution of the plant have not been satisfactory (Johnson *et al.*, 2015). Using this plant optimally could be beneficial in making a reasonable amount of micronutrients available to a large number of people.

Kale is another crop that has been in use for an extended period, and it has been acclaimed by health experts and academics for its richness in nutrients, despite having less of other vital micronutrients. It contains a significant amount of carotenoids, vitamins, minerals, prebiotic carbohydrates and minerals (Migliozzi *et al.*, 2015). Kale has been ranked high on the list of foods with a high amount of nutrients, having greater than 10% of the vital nutrients that have been thought to lower the degree of NCDs (Di Noia, 2014).

Micronutrient malnutrition affects people throughout the world. Past mechanisms to curb this tragedy emphasised crops that can be produced in massive amounts and that are easily marketed. Unfortunately, many of these crops contain limited micronutrients. Indeed, by 2011, the per capita food produced worldwide was more than 2800 kilocalories per day, a quantity that can meet the needs of every person (Migliozzi *et al.*, 2015). However, the distribution of this food is skewed, being concentrated in some regions while other regions are left with inadequate food to meet the daily dietary recommendations. Furthermore, the production of this food emphasised quantity but neglected micronutrient composition. For instance, when efforts were made to produce high-yielding foods in Bangladesh between 1980 and 1995, cereal crops such as maize, wheat, millet and sweet potatoes were produced in large quantities at the expense of vegetables, fruits and pulses, whose production reduced by over 20% (Di Noia, 2014).

According to Johnson *et al.* (2015), there have been continued efforts to address global micronutrient malnutrition. One of the principal objectives of the Millennium Development Goals crafted in 1990 was to reduce global undernutrition by 50% by 2015. These goals were partially met, as by 2013 the degree of malnutrition had decreased by 24% from the initial 15% in 1990. However, the rate of decrease in hunger efforts has fallen and therefore re-energising these goals should aim at addressing not only the quantity of food produced but also the quality (Johnson *et al.*, 2015). Thus, any initiatives addressing food inadequacy should be coupled with research to come up with affordable nutrient-rich foods in sufficient amounts.

7 Food System Approach

Nutrition, agriculture and human health are closely associated but are in many cases handled separately. An integrated strategy is critical to prevent trade-offs such as in the unregulated use of antibiotics in animal farming, which has caused ineffectiveness in the antibiotics in use in hospitals due to antibiotic resistance (Dwivedi *et al.*, 2017). Complex factors motivate food security as it involves several sectors at different levels throughout the world, and therefore focusing on the relationship between food and disease is challenging and requires multidisciplinary modelling strategies (Pingali & Sunder, 2017). In all cases, the focus should be on enhancing soil fertility, aiming at a diversity of crops incorporating fresh vegetables sustainably and affordably.

La Frano *et al.* (2014) emphasised increasing the concentration of micronutrients and their bioavailability to address the issue of obesity and other health challenges. Biofortification is a way of boosting the proportion of nutrients in commonly eaten plants by practising regulated farming whereby varieties are prearranged, as well as carrying out agronomic management and genetic engineering. Most researchers are targeting the scarcest micronutrients, particularly to some vulnerable groups like children under five years of age and pregnant women. The target micronutrients include zinc, iron, carotenoids and vitamin A (Velu *et al.*, 2014). Biofortification seems a better solution as the cost of living in many countries is high, thereby discouraging diversity in food choices, particularly for middle and low income earners. This strategy ensures that individuals consume familiar food, but with an enhanced nutrient content. Although the quantity of micronutrients in these foods may not be significantly high, most common foods are consumed in relatively large amounts, thus increasing the resultant concentration of micronutrients in the body (La Frano *et al.*, 2014).

As sustainable food production faces challenges such as global warming, there should be efforts to promote the growing of varieties of crops that endure despite climate change. Changing crop yields are associated with climate change, and therefore it is necessary to generate cultivars that are compatible with the anticipated future climate.

8 Conclusion

Although science and literature support the statement that poverty and food insecurity are fundamental constituents of the Sustainable Development Goals, statistics and research show that the current degree of malnutrition presents a challenge in realising the 2030 targets. The FAO (2019) clearly states that although economic progress is essential, this development needs to be balanced to support the reduction of poverty as well as food security globally.

Sustainable development is seeking solutions and improvements to ensure that poverty, food security and nutrition around the world are treated as equally important. Undernutrition is occurring simultaneously in developing and developed parts of the world, because while economic growth can reduce poverty and food insecurity as they are intertwined, unfortunately it will not necessarily improve the nutritional status and health of all. This is why the global and local food systems need to be shaped through sustainable development, which prioritises the health outcome of communities and individuals and especially cares about those most vulnerable in developing and developed parts of the world.

The food industry in developed countries has corrupted the eating habits of many people by promoting the mass consumption of value-added food while neglecting wholefoods, including fruits and vegetables. In developing countries, the lack of diversity in the available foods deprives the body of micronutrients, which leads to malnutrition. Both mechanisms are harmful and promote overnutrition, which can consequently cause overweight and obesity, as well as the development of obesity-related NCDs. The situation presents a double burden, dealing with hunger while at the same time addressing obesity and other lifestyle diseases.

Researchers are now focused on conventional plant breeding, an endeavour that seeks to fortify the most commonly consumed foods to raise the levels of micronutrients in these foods, particularly for targeted vulnerable groups. Government diet regulation bodies should monitor manufacturers and require them to include enough fibre in their products and to display labels showing the amount of fibre contained in them. Stakeholders should focus particularly on adolescents' health as this will affect quality of life during pregnancy and childhood. During pregnancy, appropriate micronutrient supplementation of iron, magnesium, folic acid and iodine should be carried out for the promotion of health. Public awareness concerning diet is critical and could be achieved by health promotion. The government's policies should protect against food insecurity in term of both quantity and nutritional quality by shaping a sustainable food system. Furthermore, healthcare facilities should provide sensitisation programmes for consumers to promote healthy lifestyle habits.

References

Aigner, E., Feldman, A. and Datz, C. (2014) Obesity as an emerging risk factor for iron deficiency. *Nutrients* 6 (9), 3587–3600.

Alsaffar, A.A. (2016) Sustainable diets: The interaction between food industry, nutrition, health and the environment. *Food Science and Technology International* 22 (2), 102–111.

Altieri, M.A., Funes-Monzote, F.R. and Petersen, P. (2012) Agroecologically efficient agricultural systems for smallholder farmers: Contributions to food sovereignty. *Agronomy for Sustainable Development* 32 (1), 1–13.

Andersen, M.M., Landes, X., Xiang, W. *et al.* (2015) Feasibility of new breeding techniques for organic farming. *Trends in Plant Science* 20 (7), 426–434.

Ashraf, M.Y., Ashraf, M. and Ozturk, M. (2018) Underutilised vegetables: A tool to address nutritional issues, poverty reduction and food security. In M. Ozturk, K.R. Hakeem, M. Ashraf and M.S.A. Ahmad (eds) *Global Perspectives on Underutilised Crops* (pp. 1–23). Cham: Springer.

Azab, S.F., Saleh, S.H., Elsaeed, W.F., Elshafie, M.A., Sherief, L.M. and Esh, A.M. (2014) Serum trace elements in obese Egyptian children: A case-control study. *Italian Journal of Pediatrics* 40 (1), 20.

Barbagallo, M. and Dominguez, L.J. (2015) Magnesium and type 2 diabetes. *World Journal of Diabetes* 6 (10), 1152.

Beal, T., Massiot, E., Arsenault, J.E., Smith, M.R. and Hijmans, R.J. (2017) Global trends in dietary micronutrient supplies and estimated prevalence of inadequate intakes. *PloSOne* 12 (4), e0175554.

Bhutta, Z.A., Das, J.K., Rizvi, A., Gaffey, M.F., Walker, N., Horton, S. and Maternal and Child Nutrition Study Group (2013) Evidence-based interventions for improvement of maternal and child nutrition: What can be done and at what cost? *The Lancet* 382 (9890), 452–477.

Black, R.E., Victora, C.G., Walker, S.P., Bhutta, Z.A., Christian, P., De Onis, M. and Uauy, R. (2013) Maternal and child undernutrition and overweight in low-income and middle-income countries. *The Lancet* 382 (9890), 427–451.

Carvalho, F.P. (2017) Pesticides, environment, and food safety. *Food and Energy Security* 6 (2), 48–60.

Chattopadhyay, S., Raychaudhuri, U. and Chakraborty, R. (2014) Artificial sweeteners – a review. *Journal of Food Science and Technology* 51 (4), 611–621.

Chetty, R., Stepner, M., Abraham, S., Lin, S., Scuderi, B., Turner, N. and Cutler, D. (2016) The association between income and life expectancy in the United States, 2001–2014. *JAMA* 315 (16), 1750–1766.

Clemens, R., Kranz, S., Mobley, A.R., Nicklas, T.A., Raimondi, M.P., Rodriguez, J.C. and Warshaw, H. (2012) Filling America's fiber intake gap: Summary of a roundtable to probe realistic solutions with a focus on grain-based foods. *Journal of Nutrition* 142 (7), 1390S–1401S.

Colby, S.L. and Ortman, J.M. (2015) Projections of the size and composition of the US population: 2014 to 2060. Population Estimates and Projections. Current Population Reports P25–1143. US Census Bureau. See https://files.eric.ed.gov/fulltext/ED578934.pdf (accessed 14 June 19).

Crume, T.L., Brinton, J.T., Shapiro, A., Kaar, J., Glueck, D.H., Siega-Riz, A.M. and Dabelea, D. (2016) Maternal dietary intake during pregnancy and offspring body composition: The Healthy Start Study. *American Journal of Obstetrics and Gynecology* 215 (5), 609–e1.

Di Noia, J. (2014) Peer-reviewed: Defining powerhouse fruits and vegetables: A nutrient density approach. *Preventing Chronic Disease* 11, E95.

Dwivedi, S.L., van Bueren, E.T.L., Ceccarelli, S., Grando, S., Upadhyaya, H.D. and Ortiz, R. (2017) Diversifying food systems in the pursuit of sustainable food production and healthy diets. *Trends in Plant Science* 22 (10), 842–856.

FAO (Food and Agriculture Organisation of the United Nations) (2019) *The State of Food Security and Nutrition in the World 2019.* (accessed 2 October 2019) from http://www.fao.org/state-of-food-security-nutrition/en/

Folkvord, F., Anschütz, D.J., Boyland, E., Kelly, B. and Buijzen, M. (2016) Food advertising and eating behaviour in children. *Current Opinion in Behavioral Sciences* 9, 26–31.

Gearhardt, A.N., Grilo, C.M., DiLeone, R.J., Brownell, K.D. and Potenza, M.N. (2011) Can food be addictive? Public health and policy implications. *Addiction* 106 (7), 1208–1212.

Howlett, J., Betteridge, V., Champ, M., Craig, S.S., Meheust, A. and Jones, J.M. (2010) The definition of dietary fibre – discussions at the Ninth Vahouny Fiber Symposium: Building scientific agreement. *Food & Nutrition Research* 54 (1), 5750.

Johnson, C.R., McGee, R. and Thavarajah, P. (2015) Phenotyping nutritional and antinutritional traits. In J. Kumar, A. Pratap and S. Kumar (eds) *Phenomics in Crop Plants: Trends, Options and Limitations* (pp. 223–233). New Delhi: Springer.

Johnston, J.L., Fanzo, J.C. and Cogill, B. (2014) Understanding sustainable diets: A descriptive analysis of the determinants and processes that influence diets and their impact on health, food security, and environmental sustainability. *Advances in Nutrition* 5 (4), 418–429.

Jones, A.D. and Ejeta, G. (2016) A new global agenda for nutrition and health: The importance of agriculture and food systems. *Bulletin of the World Health Organization* 94 (3), 228–229.

Joy, E.J., Green, R., Agrawal, S., Aleksandrowicz, L., Bowen, L., Kinra, S. and Dangour, A.D. (2017) Dietary patterns and non-communicable disease risk in Indian adults: Secondary analysis of Indian Migration Study data. *Public Health Nutrition* 20 (11), 1963–1972.

Katz, D.L. and Meller, S. (2014) Can we say what diet is best for health? *Annual Review of Public Health* 35, 83–103.

Ladeira, C., Carolino, E., Gomes, M.C. and Brito, M. (2017) Role of macronutrients and micronutrients in DNA damage: Results from a food frequency questionnaire. *Nutrition and Metabolic Insights* 10, 1–8.

La Frano, M.R., de Moura, F.F., Boy, E., Lönnerdal, B. and Burri, B.J. (2014) Bioavailability of iron, zinc, and provitamin A carotenoids in biofortified staple crops. *Nutrition Reviews* 72 (5), 289–307.

Leblanc, C., Dubé, M.P., Presse, N., Dumas, S., Nguyen, M., Rouleau-Mailloux, É. and Ferland, G. (2016) Avoidance of vitamin K-rich foods is common among warfarin users and translates into lower usual vitamin K intakes. *Journal of the Academy of Nutrition and Dietetics* 116 (6), 1000–1007.

Lim, S.S., Vos, T., Flaxman, A.D., Danaei, G., Shibuya, K., Adair-Rohani, H. and Aryee, M. (2012) A comparative risk assessment of burden of disease and injury attributable to 67 risk factors and risk factor clusters in 21 regions, 1990–2010: A systematic analysis for the Global Burden of Disease Study 2010. *The Lancet* 380 (9859), 2224–2260.

Mattison, J.A., Colman, R.J., Beasley, T.M., Allison, D.B., Kemnitz, J.W., Roth, G.S. and Anderson, R.M. (2017) Caloric restriction improves health and survival of rhesus monkeys. *Nature Communications* 8, 14063.

Migliozzi, M., Thavarajah, D., Thavarajah, P. and Smith, P. (2015) Lentil and kale: Complementary nutrient-rich whole food sources to combat micronutrient and calorie malnutrition. *Nutrients* 7 (11), 9285–9298.

Mozaffarian, D., Rosenberg, I. and Uauy, R. (2018) History of modern nutrition science – Implications for current research, dietary guidelines, and food policy. *British Medical Journal* 361, k2392.

Nandal, U. and Bhardwaj, R.L. (2014) The role of underutilised fruits in nutritional and economic security of tribals: A review. *Critical Reviews in Food Science and Nutrition* 54 (7), 880–890.

Nestle, M. (2013) *Food Politics*. Berkeley: University of California Press. See https://doi.org/10.1525/

Nguyen, K.H., Glantz, S.A., Palmer, C.N. and Schmidt, L.A. (2019) Tobacco industry involvement in children's sugary drinks market. *British Medical Journal* 364, l736.

Oladepo, O.I. and Abimbola, O.S. (2015) The influence of brand image and promotional mix on consumer buying decision – a study of beverage consumers in Lagos State, Nigeria. *British Journal of Marketing Studies* 3 (4), 97–109.

Oseni, G., McGee, K. and Dabalen, A. (2014) *Can Agricultural Households Farm their Way out of Poverty?* Washington, DC: World Bank.

Piernas, C., Wang, D., Du, S., Zhang, B., Wang, Z., Su, C. and Popkin, B.M. (2016) Obesity, non-communicable disease (NCD) risk factors and dietary factors among Chinese school-aged children. *Asia Pacific Journal of Clinical Nutrition* 25 (4), 826.

Pingali, P. and Sunder, N. (2017) Transitioning toward nutrition-sensitive food systems in developing countries. *Annual Review of Resource Economics* 9, 439–459.

Quann, E.E., Fulgoni, V.L. and Auestad, N. (2015) Consuming the daily recommended amounts of dairy products would reduce the prevalence of inadequate micronutrient intakes in the United States: Diet modeling study based on NHANES 2007–2010. *Nutrition Journal* 14 (1), 90.

Schlüssel, M.M., Silva, A.A.M.D., Pérez-Escamilla, R. and Kac, G. (2013) Household food insecurity and excess weight/obesity among Brazilian women and children: A life-course approach. *Cad Saude Publica* 29 (2), 219–226.

Seufert, V., Ramankutty, N. and Foley, J.A. (2012) Comparing the yields of organic and conventional agriculture. *Nature* 485 (7397), 229.

Siegel, K.R., Ali, M.K., Srinivasiah, A., Nugent, R.A. and Narayan, K.V. (2014) Do we produce enough fruits and vegetables to meet global health need? *PloSOne* 9 (8), e104059.

Singh, Y.R., Verma, S., Agrawal, D., Singh, B., Bhardwaj, A. and Agrawal, G.A. (2015) A study of magnesium supplementation on glycemic control in patients of type-2 diabetes mellitus. *Indian Journal of Clinical Anatomy and Physiology* 2 (1), 26–30.

Śliwka, U. and Przybyłowicz, K.E. (2017) The ethical and practical consideration of access to organic food as part of sustainable nutrition and human right. *EPISTEME* 36, 267–278.

Velu, G., Ortiz-Monasterio, I., Cakmak, I., Hao, Y. and Singh, R.P. (2014) Biofortification strategies to increase grain zinc and iron concentrations in wheat. *Journal of Cereal Science* 59 (3), 365–372.

Vernarelli, J.A., Mitchell, D.C., Rolls, B.J. and Hartman, T.J. (2018) Dietary energy density and obesity: How consumption patterns differ by body weight status. *European Journal of Nutrition* 57 (1), 351–361.

Waern, R.V., Cumming, R.G., Blyth, F., Naganathan, V., Allman-Farinelli, M., Le Couteur, D. and Hirani, V. (2015) Adequacy of nutritional intake among older men living in Sydney, Australia: Findings from the Concord Health and Ageing in Men Project (CHAMP). *British Journal of Nutrition* 114 (5), 812–821.

WHO (2014) *World Health Statistics 2014*. Global Health Indicators (accessed 4 February, 2019) from https://reliefweb.int/sites/reliefweb.int/files/resources/EN_WHS2014_Part3.pdf

WHO (2017) *World Health Statistics 2017*. Monitoring Health For The SDGs Sustainable Development Goals (accessed 4 April, 2019) from https://www.who.int/gho/publications/world_health_statistics/2017/EN_WHS2017_TOC.pdf?ua=1

WHO (2019) *Obesity and Overweight* (accessed 24 May, 2019) from https://www.who.int/news-room/fact-sheets/detail/obesity-and-overweight

Wimalawansa, S.A. and Wimalawansa, S.J. (2014) Agrochemical-related environmental pollution: effects on human health. *Global Journal of Biology, Agriculture and Health Sciences* 3 (3), 72–83.

Yakoob, M.Y., Salam, R.A., Khan, F.R. and Bhutta, Z.A. (2016) Vitamin D supplementation for preventing infections in children under five years of age. *Cochrane Database of Systematic Reviews* 11, CD008824.

You, D., Hug, L., Ejdemyr, S., Idele, P., Hogan, D., Mathers, C. and Alkema, L. (2015) Global, regional, and national levels and trends in under-5 mortality between 1990 and 2015, with scenario-based projections to 2030: A systematic analysis by the UN Inter-agency Group for Child Mortality Estimation. *The Lancet* 386 (10010), 2275–2286.

Zhao, L., Zhang, X., Shen, Y., Fang, X., Wang, Y. and Wang, F. (2015) Obesity and iron deficiency: A quantitative meta-analysis. *Obesity Reviews* 16 (12), 1081–1093.

5 The Internalisation of Tourism Externalities: Methods and Instruments

Agnieszka Gawlik

Contemporary tourism, a source of various benefits, also brings different kinds of costs to the economy, society and the environment. The aim of this chapter is to identify instruments for the internalisation of tourism spillover effects and to evaluate these instruments in terms of economic efficiency. This is a theoretical and empirical study. The theoretical section presents the negative and positive externalities of tourism and indicates theoretical methods of internalisation of these effects. The conceptual part of the study is an assessment of proposed internalisation instruments applied in practice with the use of the following evaluation measures: efficacy (potential and actual), cost-efficiency, distributive justice, the possibility of implementation and the existence and strength of social, political and mental barriers. The results of this assessment have practical implications for the development of tourism policies and the implementation of the concept of sustainable tourism, which is the only way to ensure the continuity and long-term development of the tourism industry.

1 Introduction

Tourism has been proven to be a significant catalyst for economic growth and job creation in many developing countries. Tourism is labour intensive and provides significant opportunities for women and unskilled workers as well as opportunities at the level of micro, small and medium-sized enterprises and communities. Therefore, tourism can play an important role as an instrument in poverty alleviation (Ashley *et al.*, 2001; Spenceley, 2008).

On the other hand, tourism may have some negative impacts on developing countries. It demands low skills and produces correspondingly low wages. It may bring disadvantages to the poor, by causing displacement, inflation, inequality and social disruption (Ashley & Roe, 2002). It may also erode the social culture and local environment in order to accommodate the demands of tourists.

The concept of **sustainable tourism** (initially known as alternative tourism, then soft tourism and environmentally sensible tourism) was a response to the expansive development of mass tourism and its negative impact on the environment and sociocultural milieu of reception regions (Collins, 1999; Kowalczyk, 2010; Xing & Dangerfield, 2011). It is related to the organisation and functioning of the tourism economy as a whole, which should ensure the long-term use of tourism assets. The UN World Tourism Organization (UNWTO, 2013) provides the following definition: 'Sustainable tourism development meets the needs of present tourists and host regions while protecting and enhancing opportunities for the future. It is envisaged as leading to management of all resources in such a way that economic, social and aesthetic needs can be fulfilled while maintaining cultural integrity, essential ecological processes, biological diversity, and life support systems.'

According to the UNWTO (2005), sustainable tourism should:

- make optimal use of environmental resources that constitute a key element in tourism development, maintaining essential ecological processes and helping to conserve natural heritage and biodiversity;
- respect the sociocultural authenticity of host communities, conserve their built and living cultural heritage and traditional values, and contribute to intercultural understanding and tolerance;
- ensure viable, long-term economic operations, providing socio-economic benefits to all stakeholders that are fairly distributed, including stable employment and income-earning opportunities and social services to host communities, and contributing to poverty alleviation.

The primary concern of sustainable tourism is to determine the acceptable level of growth in tourist traffic that does not cause negative consequences (also known as negative externalities, external costs or negative spillover effects) (Nijkamp & Verdonkschot, 1995). This involves the supply aspect (regions, tourist products, tourism enterprises and their human resources) as well as the demand aspect (sustainable consumption) (Kiryluk & Borkowska-Niszczota, 2006). Negative externalities emerge in a situation in which the activity of one entity causes a change in the welfare of another entity, and at the same time this change is not compensated (as it is outside the market mechanism) (Pigou, 1932).

2 External Effects of Tourism

Tourism has environmental, economic and social impacts. Externalities can be classified into three categories (Andereck *et al.*, 2005; Gunn, 1988; Gursoy *et al.*, 2009; Murphy, 1983; Panasiuk, 2010; Williams, 1979; Wodejko, 1998):

- **Environmental** – e.g. disorders of water balance caused by excessive water consumption and excessive emission of wastewater; oil pollution generated by water transport; pollution created by land transport or destruction of flora and fauna; landscape deterioration; soil degradation; damage to living nature (destruction of vegetation and depletion of wildlife, increased fishing, poaching); litter; land use loss.
- **Economic** – e.g. inflationary pressures; increase in operating costs of enterprises associated with increasing environmental costs; leakage of revenues; 'enclave tourism'.
- **Sociocultural** – e.g. unsatisfactory conditions of employment in tourism; the threat to the cultural identity of the local population; increased urbanisation as a result of infrastructure and spatial development of the region, leading to unfavourable conditions for the local community; increase in crime, prostitution, alcohol and drugs (Biagi *et al.*, 2012).

Externalities arising as a result of tourism development affect not only the reception areas but also areas where preparations for tourism activities are made, the zones surrounding the tourism area (emissions areas) and transit (Łoś, 2014). The pressure exerted on the reception region is the strongest; it is found in many fields. However, the impact of tourism development can also be positive and provide external benefits – they are directly related to the tourism functions. The positive impact may be: economic – local economic development, improvement of living standards, decline in unemployment, development and improvement of infrastructure and income growth (Akis *et al.*, 1996; Liu & Var, 1986; Tosun, 2002); environmental – introduction of initiatives to preserve and protect natural and artificial systems (Lindsay *et al.*, 2008); or sociocultural – development of leisure facilities and cultural exchange (Liu & Var, 1986). Sustainable tourism is about maximising the impacts that are positive and minimising the negative ones.

As the numerous examples reveal, tourism is associated with both positive and negative externalities at the same time. According to Candela *et al.* (2008), they can be defined as 'multiple externalities' or intersecting externalities in the sense that the overall externality inherent in tourism can change from positive to negative depending on the level of tourism – when, for example, the costs of congestion and degradation of nature overcome the benefits of tourism.

3 Instruments for Internalising the External Effects of Tourism

Internalisation is the inclusion of external costs or benefits arising from the activity of the entity in its economic calculations. It is a way to close the gap between the private and social costs of economic activity (Buchanan, 1969; Ethridge, 1973; Narver, 2017; Puciato & Łoś, 2009).

Policy instruments for sustainable tourism management are no different in essence from instruments in other fields of environmental policy. They can be divided into three groups: economic (market-based), regulatory (command-and-control) and institutional (OECD, 1991). Economic instruments comprise tourist taxes, user fees, financial incentives and tradable market permits. Regulatory instruments include quotas and zoning. Institutional instruments refer to eco-labels and changes in property rights Most commonly, a combination of various policy instruments is used.

Particular instruments differ in the way in which they internalise externalities. Their task is to assign the external costs of tourism, compensation costs of environmental damage and loss, costs of pressure reduction in reception areas and costs of prevention.

Economic instruments (also called market-based instruments – MBIs) are basically market mechanisms that are designed to address context-specific problems of environmental quality including natural and cultural resources for tourism. Tourist taxes, user fees and financial incentives are used to change the behaviours of users so they are aligned with the objectives of sustainable management (Stavins, 2001).

From the economics literature, it is well known that externalities can be internalised by using **taxes** (Pigouvian tax) or **subsidies.** It is therefore straightforward to correct for the externalities associated with tourism by imposing a tourism tax (or subsidy in the case of a positive overall externality). The objective of tourism taxes is twofold. Firstly, it is to generate revenue and, as already indicated, to correct for market failures. The second purpose of taxation, the internalisation of external effects, aims at achieving a socially desirable outcome. This is an important issue, because if such an intervention in markets is properly done, overall welfare in the host country or region improves, thus increasing residents' quality of life. The issue of optimal tourism taxation has been the subject of various studies (Cerina, 2007; Gomez *et al.*, 2008; Piga, 2003; Pintassilgo & Silva, 2007). The studies indicate that with proper taxation, negative external effects can be prevented by influencing the market. On the other hand, there are examples in which the introduction of tourism taxation is purely and simply an instrument designed to generate revenue (e.g. Balearic islands).

Financial incentives for small businesses can be an instrument that develops tourism from the bottom up, according to the idea of '**community-based tourism**' or '**pro-poor tourism**'. Small and medium-sized enterprises (SMEs) present greater opportunities for control and profit by local people. Guesthouses import less than hotels and, as a result, the multiplier is greater (Milne, 1992). Furthermore, small-sized tourism enterprises have a greater impact on improving rural living standards, reducing rural–urban migration and countering structural inequalities of income distribution (Britton, 1982). Small-scale tourism places value on natural and cultural resources and can be developed without great capital investment. As tourism can stimulate employment, the drive for young people to migrate to urban areas is reduced, leaving a more balanced population in remote rural areas.

Very often, fees and taxes are singled out as one group of instruments. The difference between the two lies mainly in the direction of spending funds collected by fees. **Tourism user fees** gather revenue from tourism-based activities which can be directed towards restoring and maintaining valuable tourism resources (Dwyer *et al.*, 2010). The fees partially reflect the cost of supplying recreational services, the demand for natural resources and the value that visitors attribute to the reception area. The direct link between maintaining natural areas and income from user fees is a strong economic incentive for conservation (CFA, 2004).

The tools that are obligatory form a group of administrative and legal methods. Instruments that arise from autonomous decisions of companies cannot be included in this group. Prerequisites for their use are, most commonly, sustainable tourism management awareness and the demand of consumers.

Eco-labelling is a voluntary system of awarding a logo to products or services judged to be less harmful to the environment than comparable ones. Eco-labelling is used all over the world. Eco-labels inform consumers about a specific product (service) and are granted by a third party to the particular kinds of goods and services for which environmental criteria are determined independently. Eco-labelling can help tourism suppliers identify critical environmental issues and encourage the implementation of eco-efficient solutions. Eco-labels identify products and services that can decrease the use of resources, such as energy or water, and reduce costs for the operator. This instrument can play a direct role in linking socioeconomic concerns with environmental protection. For example, eco-labelling may encourage the use of more regional products, benefitting the local economy and reducing the need for transport. It can also help to raise awareness of environmental protection among the regional and local population as well as highlighting the importance of tourism for the environment in general. An example of the introduction of a simple labelling technique in accommodation or catering services is the island of Hiiumaa (Estonia) in 1996.

Quotas set fixed maximum numbers of visitors to be admitted to destinations during given periods. They may involve: closure of certain places, such as environmentally fragile areas, at certain times; establishing a maximum number of accommodation units; or determining a maximum number of persons allowed at a certain tourist attraction, a particular area or a whole country. Quotas prevent overcrowding and, consequently, natural resource degradation. In 1974 Bhutan became the only country to have introduced a tourist quota at the national level (although this quota has now been removed, and as a result the number of tourists increased) (Brunet *et al.*, 2001).

Zoning regulation can be a very effective instrument for limiting construction activity, which is one of the biggest problems related to environmental degradation caused by tourism development. This instrument is relatively inexpensive and easy to implement. The basic implementation document is a physical plan. It can restrict construction in environmentally sensitive areas (e.g. 100 metres from the coast) or minimise areas allocated for new construction activities (e.g. in the Maldives, regulations state that the built environment should utilise no more than 20% of the total land area). It can also determine development standards (building density, height limits) that impact the layout and design of tourist facilities.

A summary of the main characteristics of particular groups of instruments and their examples is presented in Table 5.1.

Table 5.1 Main instruments for the internalisation of tourism externalities

No.	Type of instrument	Description	Examples
1	Tourist tax	Taxes that are paid by the majority of tourists for staying in a tourism area	Arrival/departure tax, bed tax, occupancy tax, accommodation tax, service tax, eco-tourism tax, visa and travel permits, air passenger duties
2	User fees	Fees charged to tourists in exchange for access to specific resources	Entrance fees to protected areas
3	Financial incentives	Subsidies for operators to cover (fully or partially) the cost of installing devices or actions to reduce the external costs created	Lower taxes, subsidies, reduced import tariffs encouraging environmentally friendly equipment for water and energy saving at hotels
4	Eco-labels	Labels applied to products or services offered to tourists that satisfy certain environmental criteria	Eco-labelling on products and services such as accommodation facilities, tour operators, beaches, restaurants, marinas or tourist destinations (e.g. Blue Flag, Green Globe 21)
5	Quotas	Limits on the numbers of visitors admitted to destinations during fixed periods	A maximum number of accommodation units; a maximum number of persons allowed at a certain tourist attraction, particular area or a whole country (e.g. Bhutan)
6	Zoning	Limits on construction activities	Restrictions on construction in environmentally sensitive areas (e.g. 100 metres from the coast), minimising areas allocated for new construction, development standards – building density and height limits

Source: Author's own analysis.

4 Measures for Evaluating Internalisation Instruments in Sustainable Tourism Policy

The design of these instruments should be guided by the principle that the sum of social benefits gained by reducing the extent of loss in welfare should be higher than the total costs associated with the implementation and functioning of the instrument. It is necessary, however, to establish criteria that enable relatively quick evaluation of considered instruments.

The basic measures for evaluating internalisation instruments should include:

- **Efficacy (potential and actual)** – the degree of achievement (potential and actual) of the objectives of sustainable tourism. This depends on the implementation, enforceability and impact on the economic efficiency of the chosen solution. Improving the efficacy of instruments should proceed in two directions. First of all, existing instruments that have proven their efficacy must be used and improved. Secondly, new instruments should be introduced, based on the experiences of other countries, with efficacy proven in practice.
- **Cost-efficiency** – the relationship between results (B – benefits) and the expenditure (C – costs); it should be greater than one. The cost-efficiency criterion is usually understood as:
  - minimising the total cost of achieving the planned, given benefit or minimising the cost of obtaining a unitary effect;
  - the existence of a surplus of income generated by a given instrument over the costs of its implementation and its current functioning (sometimes referred to as administrative efficiency).
  - The first approach is called 'external efficiency' and the second, 'internal efficiency' of the operating system of a given instrument.
- **Distributive justice** – all fees, taxes and other charges directly or indirectly affect the prices of goods and production factors. This leads to distributional effects that involve enterprises, households and countries. Distribution of these effects should be socially equitable. In terms of 'pro-poor tourism', it means that the access of the poor to economic benefits should be increased (expanding business and employment opportunities for the poor and providing training, for example). Moreover, negative social and environmental impacts associated with tourism (such as lost access to land, coastal areas and other resources, social disruption or exploitation) should be addressed.
- **The possibility of implementation** – the existence of legal and institutional foundations; the existence of organisational and technical solutions for the introduction and supervision of the instrument; and social and political acceptability of the instrument, which is the result of the attitude of many groups of enterprises, consumers,

non-governmental tourist and environmental organisations, institutions and so on. The first two groups of factors always involve specific costs and are most often of an administrative and legal nature, although there are also barriers to implementation resulting from conceptual problems related to the design of institutional and organisational solutions necessary for the implementation of a given instrument.

- **The existence and strength of social, political and mental barriers** – acceptability is higher if there are precisely defined groups (economic, social) affected by the impact of the instrument; information about the nature of the instrument, its implementation and effects is widespread; implementation of the instrument is gradual and meets the condition of informing groups covered by the effects of the instrument in advance about the intention to move to the next implementation stages.

Additional criteria may include the possibility of identification of the external effect and the existence of appropriate information necessary for implementation of the instrument. When designing instruments for internalising external tourism costs, it is important to identify the possibility of recognising the problem and the existence of appropriate information necessary to prepare the implementation project. The essence of the problem should be recognised based on knowledge from natural, technical and social sciences. The health and environmental risks should be determined, as well as the deterioration in the quality of life that it generates and the availability of current and reliable data. If these actions are not taken, the instrument may not function properly (especially in terms of cost-efficiency and efficacy), which may make the instrument uneconomic and/or pointless.

Table 5.2 shows the example of using a set of evaluation criteria for comparing different instruments for internalisation in mountain tourism. The purpose of these instruments was to reduce human pressure on mountain trails. Mountain trails in Tatra National Park in Poland were used as an example. In 2018 nearly 3.8 million people entered Tatra National Park (TPN, 2018). Interest in the park, which covers 212km^2, is growing rapidly. In 2007, the number of tourists was 2.7 million. In 2013 it was just under 3 million (TPN, 2013, 2017). According to research, 70%–80% of tourists most often choose the most popular tourist destinations: the Morskie Oko, Kasprowy Wierch (including entry by cable car), Kościeliska Valley in the Western Tatras or Giewont. These tourist attractions are adding to the problems of overcrowded car parks near the entrances to trails, crowding and long queues (e.g. the Kasprowy queue or the queue to chains on Giewont). Human pressure in the Tatra mountains is greatest at Kościeliska Valley in the Western Tatras and Giewont. Tourists complain about the crowds on the trails, and the trails have a limited capacity.

Table 5.2 Assessment of instruments to reduce pressure from tourism on the Tatra mountain trails

Criteria \ Instrument	Pro-Mountains Eco-tax	Higher entrance fees on mountain trails	Mountain trail quotas
Efficacy	Low (1) – revenue-raising instrument, weakly deters entry	High (4) – charges raised allocated to environment protection or reparation	Good (3) – limits the number of tourists
Cost-efficiency	Good (3) – administrative costs can be high	Good (3) – charges raised allocated to environment protection or reparation	Good (3) – administrative costs can be high
Distributive justice	Sufficient (2) – lower income groups priced out	Good (3) – lower income groups receive discounts	Sufficient (2) – limited economic opportunities for the local community
Ability to implement	Good (3) – other taxes are already present, easy to monitor	Sufficient (2) – fees are already present, hard to monitor	Low (1) – need for appropriate legal regulations, very hard to monitor
The existence and strength of barriers	Good (3) – low social barrier	Sufficient (2) – social barrier	Low (1) – high social and political barriers
Total score	12	14	10

Source: Author's own analysis.

The evaluation covered three proposed instruments: a special Pro-Mountains Eco-tax, an increase to the existing entrance fees on all/the most popular trails, and mountain trail quotas. A simplified rating scale was adopted: 4 – high; 3 – good; 2 – sufficient; 1 – low.

This simplified assessment with evaluation criteria demonstrates that the best instrument, in this case, is the raising of existing fees for access to the trails (especially the most popular ones). The high economic efficiency of this instrument is mainly the result of the fact that charges raised are allocated to environment protection or reparation. This is the implementation of the 'polluter pays principle', which enables the realisation of sustainable tourism in practice. Raising of existing fees could reduce tourist traffic in the most frequented places. It must be mentioned that alongside the implementation of this instrument, tourists should be informed about the current traffic volume in a given place and other tourist attractions located in the region. This would require a long-term horizon of activities involving the development of a coherent strategy for the sustainable development of tourism in the entire region.

Most commonly a combination of different instruments is used to achieve the objectives. It is important that they form a coherent policy which contributes to the implementation of sustainable development.

Table 5.3 Evaluation criteria at the stages of formulation, realisation and identification of the impact of sustainable tourism policy

Stage of the process	Criteria	Proposed assessment measures
I. Formulation	Efficacy (potential) The existence and strength of social, political and mental barriers The ability to implement	1. The opinions of the local community, politicians, administrators and other actors involved in the process 2. The scope of the necessary legal changes, the size and structure of the administrative apparatus
II. Realisation	The existence and strength of social, political and mental barriers The ability to implement	1. The impact of interest groups 2. The availability of monitoring techniques 3. The ability to mitigate the adverse effects of changes
III. Identification of impact	Efficacy (actual) Cost-efficiency Distributive justice	1. Change in behaviour of tourists and tourism enterprises 2. Quality of life indicators of the local community 3. Degree of achievement of objectives 4. Pricing distribution

Source: Author's own analysis.

What is more, local communities should be given the maximum opportunity to participate in tourism development and decision-making processes surrounding tourism, to improve their livelihoods by securing better access to tourists and tourism enterprises. Therefore, the chosen instruments should encourage partnerships between the private sector and poor people. For instance, this should occur during the development of new tourism products.

Table 5.3 presents the use of evaluation criteria at the stages of formulation, realisation and identification of the impact to form a coherent sustainable tourism policy. Some assessment measures for the criteria were proposed. It should be noted that specific measures can be customised to the needs of decision makers.

5 Conclusion

The intensification of international exchange, technological progress, globalisation and changes in the lifestyles of modern people mean that tourism activity is becoming an increasingly common way to spend leisure time. The consequence of these trends is the dynamic development of tourism, which is increasingly important for the functioning of many national economies. However, the impact of tourism does not end at the economic level but also affects the social, spatial, cultural and environmental spheres. The development of tourism is not possible without a natural environment, which functions as its basis. Therefore, it is extremely important to take appropriate action, the essence of which is to prevent the negative external effects of tourism.

The negative impacts of tourism should be internalised by using appropriate methods and instruments. Internalisation enables the achievement of sustainable tourism in practice. However, instruments for internalisation differ in economic efficiency, and sustainability in tourism requires their assessment. Evaluation measures should include efficacy (potential and actual), cost-efficiency, distributive justice, the possibility of implementation, and the existence and strength of social, political and mental barriers.

The proposed evaluation criteria enable the assessment and comparison of instruments. Firstly, they allow the use and improvement of existing instruments that have been confirmed to be effective. Secondly, they allow for the evaluation of new instruments that have proved effective in other countries.

The list of proposed evaluation criteria should be used in all stages of the formulation, realisation and identification of the impact of sustainable tourism policy. This will ensure that the policy is coherent and efficient.

References

Akis, S., Peristanis, N. and Warner, J. (1996) Residents' attitudes to tourism development: The case of Cyprus. *Tourism Management* 17 (7), 481–494.

Andereck, K.L., Knopf, R.C. and Vogt, A. (2005) Residents' perceptions of community tourism impacts. *Annals of Tourism Research* 32, 1056–1076.

Ashley, C. and Roe, D. (2002) Making tourism work for the poor: Strategies and challenges in southern Africa. *Development Southern Africa* 19 (1), 61–82.

Ashley, C., Roe, D. and Goodwin, H. (2001) *Pro-poor Tourism Strategies: Making Tourism Work for the Poor. A Review of Experience.* London: Overseas Development Institute.

Biagi, B., Brandano, M.G. and Detotto, C. (2012) The effect of tourism on crime in Italy: A dynamic panel approach. *Economics: The Open-Access, Open-Assessment E-Journal* 6, 2012–2025.

Britton, S. (1982) The political economy of tourism in the third world. *Annals of Tourism Research* 9, 331–358.

Brunet, S., Bauer, J., DeLacy, T. and Tshering, K. (2001) Tourism development in Bhutan: Tensions between tradition and modernity. *Journal of Sustainable Tourism* 9 (3), 243–263.

Buchanan, J.M. (1969) External diseconomies, corrective taxes, and market structure. *American Economic Review* 59 (1), 174–177.

Candela, G., Castellani, M. and Dieci, R. (2008) Economics of externalities and public policy. *International Economic Review* 55, 285.

Cerina, F. (2007) Tourism specialization and environmental sustainability in a dynamic economy. *Tourism Economics* 13 (4), 553–582.

CFA (Conservation Finance Alliance) (2004) *Sustainable Finance for Protected Areas: Tourism-based User Fees.* See https://www.researchgate.net/publication/256013408_Sustainable_Finance_of_Protected_Areas_Tourism_Based_User_Fees_Conservation_Finance_Guide_Conservation_Finance_Alliance.

Collins, A. (1999) Tourism development and natural capital. *Annals of Tourism Research* 26 (1), 98–109.

Dwyer, L., Forsyth, P. and Dwyer, W. (2010) *Tourism Economics and Policy* (1st edn). Bristol: Channel View Publications.

Ethridge, D. (1973) The inclusion of wastes in the theory of the firm. *Journal of Political Economy* 81 (6), 1430–1441.

Gomez, C.M., Lozano, J. and Rey-Maquieira, J. (2008) Environmental policy and long-term welfare in a tourism economy. *Spanish Economic Review* 10 (1), 41–62.

Gunn, C.A. (1988) *Tourism Planning*. New York: Taylor & Francis.

Gursoy, D., Chi, C.G. and Dyer, P.K. (2009) An examination of locals' attitudes. *Annals of Tourism Research* 36 (4), 723–726.

Kiryluk, H. and Borkowska-Niszczota, M. (2006) Koncepcja turystyki zrównoważonej w teorii i praktyce. In A. Nowakowska and M. Przydział (eds) *Turystyka w badaniach naukowych* (p. 111). Rzeszów: Wyższa Szkoła Informatyki i Zarządzania w Rzeszowie.

Kowalczyk, A. (ed.) (2010) *Turystyka zrównoważona* (p. 21). Warsaw: PWN.

Lindsay, K., Craig, J. and Low, M. (2008) Tourism and conservation: The effects of track proximity on avian reproductive success and nest selection in an open sanctuary. *Tourism Management* 29 (4), 730–739.

Liu, J.C. and Var, T. (1986) Resident attitudes toward tourism impacts in Hawaii. *Annals of Tourism Research* 13 (2), 193–214.

Łoś, A. (2014) Pomiar efektywności działalności gospodarczej w turystyce. *Ekonomika i Organizacja Przedsiębiorstwa* 1 (768), 30–39.

Milne, S. (1992) Tourism and development in south Pacific microstates. *Annals of Tourism Research* 19, 191–212.

Murphy, P.E. (1983) Perceptions and attitudes of decision-making groups in tourism centers. *Journal of Travel Research* 21 (3), 8–12.

Narver, J.C. (2017) Rational management responses to external effects. *Academy of Management Journal* 14, 1.

Nijkamp, P. and Verdonkschot, S. (1995) *Sustainable Tourism Development: A Case Study of Lesbos*. Amsterdam: VU University, Faculty of Economics, Business Administration and Econometrics.

OECD (1991) *Guidelines for the Application of Economic Instruments In Environmental Policy*. Paris: Organisation for Economic Co-operation and Development.

Panasiuk, A. (2010) *Ekonomika turystyki i rekreacji*. Warsaw: PWN.

Piga, C.A.G. (2003) Pigouvian taxation in tourism. *Environmental & Resource Economics* 26 (3), 343–359.

Pigou, A. (1932) *The Economics of Welfare*. London: Macmillan

Pintassilgo, P. and Silva, J. (2007) 'Tragedy of the commons' in the tourism accommodation industry. *Tourism Economics* 13 (2), 209–224.

Puciato, D. and Łoś, A. (2009) *Wykorzystanie opłat ekologicznych jako instrumentu zarządzania i finansowania pośredniej gospodarki turystycznej w obszarze ochrony środowiska*. Studia i Monografie Politechniki Opolskiej no. 252. Opole: OW Politechniki Opolskiej.

Spenceley, A. (2008) *Responsible Tourism: Critical Issues for Conservation and Development*. London: Earthscan.

Stavins, R. (2001) Experience with market-based environmental policy instruments. Discussion Paper No. 1–58. Resources for the Future. See http//www.rff.org (accessed 1 June 2019).

Tosun, C. (2002) Host perceptions of impacts: A comparative tourism study. *Annals of Tourism Research* 29 (1), 231–253.

TPN (Tatrzański Park Narodowy) (2013) *Zestawienie sprzedaży biletów 2013*. See https://tpn.pl/zwiedzaj/turystyka/statystyka (accessed 1 June 2019).

TPN (Tatrzański Park Narodowy) (2017) *Zestawienie sprzedaży biletów 2017*. See https://tpn.pl/zwiedzaj/turystyka/statystyka (accessed 1 June 2019).

TPN (Tatrzański Park Narodowy) (2018) *Zestawienie sprzedaży biletów 2018*. See https://tpn.pl/zwiedzaj/turystyka/statystyka (accessed 1 June 2019).
UNWTO (UN World Tourism Organization) (2005) *Making Tourism More Sustainable – A Guide for Policy Makers*. Nairobi and Madrid: UNEP and UNWTO.
UNWTO (UN World Tourism Organization) (2013) *Sustainable Tourism for Development Guidebook*. See https://www.e-unwto.org/doi/book/10.18111/9789284415496.
Williams, T.A. (1979) Impact of domestic tourism in host population: The evolution of a model. *Tourism Recreation Research* 4 (2), 15–21.
Wodejko, S. (1998) *Ekonomiczne zagadnienia turystyki*. Warsaw: WSHiP.
Xing, Y. and Dangerfield, B. (2011) Modelling the sustainability of mass tourism in island tourist economies. *Journal of the Operational Research Society* 62 (9), 1742–1752.

6 Slow Tourism: A Community-based Approach for Sustainable Development

Ozan Kaya

Developments that emerged after the Industrial Revolution turned tourism into a leisure activity accessible to the larger masses. In parallel to this, tourism has grown at a speed that most sectors have not been able to match. However, the rapid growth of tourism has caused various problems in social, economic and ecological dimensions. Slow tourism, which emerged as an alternative to the dominant paradigm in tourism, is a sustainable, local and community-centred approach to tourism. In contrast to mainstream tourism, slow tourism is community based and oriented towards local development. In addition, it provides methods for the sustainable integration of tourists and tourism within destinations and communities. With this in mind, the main objective of this chapter is to demonstrate how slow tourism can achieve sustainable development. The findings of the literature review were used to help determine this aim. The literature makes clear that slow tourism has great potential for sustainable development and poverty reduction through regular income distribution, low-carbon travel and accommodation, and practices that ensure and support the use of local production and agriculture.

1 Introduction

The Industrial Revolution and subsequent changes affected not only production systems, but also social structures, businesses and methods of work, as well as daily life practices. In particular, the neoliberal policies that emerged after World War II and gained momentum in the 1980s have affected the whole world and have led to many debates on the consequences of globalisation to date. One of the main criticisms of globalisation is the largely unilateral realisation of the kind of freedom and diffusion that the term globalisation implies. While in an economic, political, cultural and social context First World countries can easily shape Third World countries, the opposite is not the case (Britton, 1982).

Today, the most important pillar of globalisation is economic. Opponents of globalisation have criticised neoliberal globalisation in the economy because it increases inequalities, causes new kinds of hegemony, is ruled by the powerful, operates for the benefit of the minority, obstructs justice and destroys democracy. Howard (2012) argues that the business mindset that emerged after the Industrial Revolution forms the basis of the problems now faced in social and ecological environments. The mass production of the Industrial Revolution was reflected within the tourism industry by mass tourism. Package tours, introduced by Thomas Cook, increased rapidly following World War II. Over the last 60 years, tourist expenditure and the number of international tourists have increased remarkably. In particular, the economic opportunities provided by this increase have led to the inclusion of tourism on the agenda of both developed and developing countries. The extraordinary increase in tourism in a short space of time, however, means that destinations and local communities now face significant environmental, sociocultural and economic problems. Most researchers state that these problems are caused by mass tourism, which has increased rapidly alongside globalisation (Hall & Brown, 2006; Molz, 2009).

The negative impacts of mass tourism include the consumption of limited natural resources, tourism overinvestments and flows without consideration for carrying capacity and social erosion, with some destinations having to accommodate visitor numbers of more than 10 times the local population (Archer *et al.*, 2005; Capocchi *et al.*, 2019). Issues such as these resulted in the closure of some of the most important beaches in Thailand (Ellis-Petersen, 2018), and in protests against tourism from local residents, as seen in Barcelona (UNWTO, 2018a). In the meantime, the economic benefits of mass tourism were shared by local elites or foreign investors, with only minimal benefits for low-income locals. In tourism regions where the difference in income distribution increases, locals who work for low wages must cope with problems such as increasing rental and living costs because of regional inflation due to tourism (Ramukumba, 2015; Rettinger & Apollo, 2017).

One of the new approaches to remedy the current and potential problems arising from mass tourism is called slow tourism, which is based on the Slow Food movement. In this context, slow tourism is a form of tourism that satisfies the basic concepts of 'good', 'clean' and 'fair', in which priority should be given to localisation, sharing and sustainability (Oh *et al.*, 2014; Yurtseven *et al.*, 2010). Slow Food and slow tourism emerged as reactions to the problems caused by mass tourism, both locally and globally (Dickinson, 2009). In contrast to mass tourism, slow tourism is community based and oriented towards local development. It provides methods for the sustainable integration of tourists and tourism within destinations and communities (Molz, 2009). With this in mind, the main objective of this chapter is to demonstrate how slow tourism can achieve sustainable development goals.

2 Tourism, but How?

In the early stages of mass tourism, which could be seen as a reflection of the Fordist production approach that emerged after the Industrial Revolution, it was approached with optimism (Butcher, 2017). In this context, the positive economic effects on the balance of payments, employment and foreign exchange inflow have turned optimism for mass tourism in developing countries to interest and sympathy. This sympathy and interest was reflected in the number of international tourists and increase in international tourism revenue in the following years. The number of international tourists increased from 25 million in 1950 to 165 million in 1970, 435 million in 1990, 950 million in 2010 and 1.4 billion in 2018. Similarly, international tourism revenue, which was US$2 million in 1950, reached US$17 billion in 1970, US$271 billion in 1990, US$986 billion in 2010 and US$1.45 trillion in 2018 (UNWTO, 2018b, 2019).

Tourism, which has various positive economic effects, can, upon a closer look, also be seen to bring or increase different economic problems. Britton (1982) attributes this to the fact that discussions about the advantages or disadvantages of tourism are made without addressing the increasing inequality or poverty issues between countries. Furthermore, Scheyvens and Hughes (2019) state that in analysis of the economic effects of tourism, focus should be given to the sociopolitical aspects of poverty rather than economic growth, job and income-generating effects, as well as how structural inequalities affect sustainable development. For instance, although economic equality is a controversial issue in societies, it is an important part of sustainable development (Lee, 2009). Wen and Tisdell (1997) share findings from demand and supply analysis that tourism in various regions of China increased regional inequality. Similar results are presented by Incera and Fernandez (2015). Accordingly, in the context of tourism income, high-income households benefit more than low-income households, which contributes to an increase in income inequality in the region. Even in the United States, it is seen that regions dependent on tourism services have more income inequality than districts connected to non-tourism services. In addition, inequality in regions linked to tourism services has increased more rapidly than in the country as a whole (Lee, 2009).

Regional disparities and differences in income distribution can also be caused by market structure. An oligopoly market structure in destinations also increases inequality between the rich and the poor. The creation of basic tourism supply resources by several multinational companies results in small and medium-sized companies not being able to compete with large companies, and unfair distribution of income (Alam & Paramati, 2016). Social welfare has been pushed into the background of the distribution of tourism revenue due to unequal trade relations among countries and regional power structures at destinations that shape the exchanges between tourism-generating countries and destinations (Bianchi, 2017). According to Brohman (1996), the reason why international mass tourism cannot contribute to the economies of developing countries is that global enterprises provide monopolistic control of ownership and organisational structure in mass tourism in most countries.

Ramukumba (2015) summarises the development of tourism in a region, the transition process to mass tourism and its results as follows:

> As a new nation or region is initially incorporated into the global 'patch-work- quilt' of tourism destinations the emergent industry is characterized by relatively high levels of local involvement. As visitor numbers rise and incorporation into the global tourism system increases, local industry structures soon become characterized by overseas or local élite ownership. Locals end up receiving a few economic benefits while having to carry the inevitable costs of rapidly increasing tourist numbers. Unfortunately, the high costs associated with the inevitable rise of mass tourism provide both the communities and localities that rely on it with little prospect for local control and limited potential to achieve more sustainable forms of local development. (Ramukumba, 2015: 52)

The uncontrolled growth experienced in tourism for the last 60 years has caused significant problems at destinations. Today, tourism is one of the major threats to the environmental integrity of host regions in many parts of the world. Arıkan *et al.* (2016: 222) named the situation a tourism paradox, 'where tourism industry destroys the natural and cultural environment that is necessary for tourism activities'. This may be due to the belief in the existence of nature for the benefit of humans in the dominant Western tradition (Singer, 1993), or due to the mentality of Toffler's (1980) Second Wave (Industrial Phase), which sees nature only as an object. In the 1970s, when the interaction between various sectors and the environment was discussed, the negative impacts of tourism on the environment became more apparent due to its growth (Holden, 2008). The first international meeting on the tourism–environment relationship was the Manila Tourism Summit in 1980, which addressed the impact of the natural and cultural environment on tourism development and demonstrated that environmental values are essential for the long-term benefit of tourism (European University Centre for Cultural Heritage, 2018). According to Edgell (2016), the Manila Tourism Summit was very important in terms of providing a basis for discussions about sustainable tourism before the 1987 Brundtland Report, and led to a more basic policy document than the 1992 United Nations Conference on Environment and Development (Agenda 21) in terms of sustainability.

In this aspect, the ecological environment is one of the most important elements of tourism. However, the relationship between tourism and the ecological environment largely takes the form of excessive use and destruction of ecological resources in order to meet people's needs and requests (Edgell, 2016; Holden, 2008).

Most researchers (Becken *et al.*, 2001; Filimonau *et al.*, 2014) have raised concerns regarding global environmental impacts, such as the impact of tourism on global warming and climate change. Lenzen *et al.* (2018) report that, between 2009 and 2013, the global carbon footprint of tourism has increased four times more than previously estimated, accounting for about 8% of global greenhouse gas emissions. Similarly, in 2017, luxury cruise brands emitted ten times more sulphur oxides in European seas than 260+ million passenger vehicles in Europe (Abbasov *et al.*, 2019). Furthermore, when taking into account the destruction of local ecological balance caused by tourism, a more distressing picture is seen (see Table 6.1). To illustrate this point, the amount of water consumed by a single tourist varies from 84 litres to 2000 litres per day (Gössling *et al.*, 2012). Many factors – such as the geographical location, the accommodation rating and the facilities offered by the hotels – influence consumption amounts (Bruns-Smith *et al.*, 2015). For example, a swimming pool in a hotel increases tourism-related average water consumption by 60 litres per person. An average golf course consumes 1 million cubic metres of water per year, which is equal to the annual water consumption of approximately 12,000 people (Schouten, 2003). Although water consumption figures vary, the amount of water consumed by tourists in island destinations, such as Cyprus, reaches up to 16.9% of total

Table 6.1 Environmental impacts of tourism

Impacted field	Basic impacts
Air	Air pollution caused by transportation and tourism enterprises
Water	Problems caused by wastewater on water resources and excessive use of water resources
Geology and soil	Physical and chemical changes in soil due to collection, vandalism and waste caused by tourism. Destruction of agricultural land due to excessive land use in tourism
Landscape	Visual changes in settlements with the development of tourism. Extinction of local architecture
Habitats	Decrease of natural habitats for tourism activities and establishment of tourism facilities; disappearance of fragile species and replacement by more resilient species, changes in species composition and disappearance of rare species
Wildlife	Negative effects of environmental pollution, overfishing, fishing, highways and hiking paths on wildlife areas; changes in species composition and social behaviour; disappearance of rare species and reduction of habituated animals

Source: Spenceley (2005: 138–139).

consumption, creating significant pressure on limited water resources (Kotios *et al.*, 2013).

As tourism becomes a mass event, interactions between tourists and local people cause significant social influences and changes. Yasumura (1994) states that the relationship and interaction process between tourist and resident generally occurs in three areas. These are: (1) processes in the trade of goods and services; (2) encounters/interactions in recreation areas; and (3) interactions during the exchange of information. In this context, the most common relationship type in touristic activities occurs during the exchange of goods and services in commercial areas (Yasumura, 1994). According to Paul (2012), it is not only individuals who are encountered through tourism, but also their values, attitudes, behaviours, personalities and social cultures.

In all of these relations between tourists and local people, the individual is both an affected and an affecting element. However, in the hegemonic structure of mass tourism, the local population is the affected party while the tourists form the affecting element (Kaya, 2016). As Lea (2001) states, in the process of sociocultural interaction between tourists and local people, cultural diffusion is mainly directed towards local culture from tourist culture. As a result, negative sociocultural impacts can occur, especially in developing regions, including an increase in prostitution, gambling, crime and alcohol and drug use due to the weakening or destruction of local culture, identity and traditions, and changes in social cohesion and morality (Archer *et al.*, 2005; Ghaderi, 2013; Postma, 2013; Zhuang *et al.*, 2019).

Urry (2002) indicates that while the massification of tourism provides the democratisation of tourism, on the other hand, the regions that accept tourists have become consumption centres for both tourists and local people. Furthermore, performances and demonstrations become features of the social structure in almost all destinations. Accordingly, tourism activities that take place without considering the balance of protection lead the tourism stakeholders to new quests for both structural and operational aspects in order to achieve sustainable development goals.

3 The Emergence of the Slow Movement

Perhaps, as Milan Kundera noted in his novel *Slowness* (1997), speed is a kind of ecstasy given to humanity by the technological revolution. The short-term happiness offered by this ecstasy has created the perception that speed is mostly positive. That is why, for example, to keep up with the modern world, people have started to sleep less and take less paid holiday leave, and they use indirect methods of communication such as Post-it notes, short message system (SMS) and email, even with their family members (Honore, 2004). The passion for speed in people's personal lives is a reflection of the global business mentality of people's working lives (Knox, 2005).

The societal paradigm of speed that started in the Western world with the Industrial Revolution has now affected the entire world, and speed has become a basic element in our lives. Increasingly, time is seen as an important cost element in the global economic system. For example, many of us want faster computers, internet connections and means of transportation. Yet, at the same time, we want more balanced lives. The University of California, Los Angeles (UCLA) psychiatrist Peter Whybrow at the Semel Institute for Neuroscience and Human Behavior stated that the physical and psychological limits of many individuals are being strained under time pressures (Cecile, 2006). Moreover, they are not the only people whose limits are enforced under the speed regime. Economic growth, expansion of production and the increase in the speed of daily living show that the world is moving towards overburdening its carrying capacity (Heinonen *et al.*, 2006).

In the last quarter of the 20th century, the world has experienced an unprecedented degree of transformation at economic, political and cultural levels. Globalisation under the leadership of multinational corporations has changed production relations and has given capital flows and the market economy a global dimension by downplaying the borders between nation-states. This has created global companies that have spread their products all over the world. Global brands such as McDonald's, Starbucks, Coca-Cola and Levi Strauss & Co. have become symbols of the regime of speed, and they have become a threat to local products and local cultures. Lipschutz (2002) states that the main objective of this hegemonic paradigm is to strive to transform the existing societies into societies that are disciplined and homogeneous and that believe that they can be differentiated by consuming.

In 1986, Carlo Petrini protested the opening of a fast food restaurant (McDonald's) in Rome and started the Slow Food movement. Slow Food is a civil society movement that aims to protect ethnic and local cuisine against the threats posed by the global world (Slow Food, 2019). The Slow Food movement was officially established in 1989 at the international Slow Food Congress in Paris and the Slow Food Manifesto was signed. The Manifesto presents a systematic critique of the 'fast life', which is rooted in the contemporary world. It also demonstrates the need for a holistic alternative in a culture that is not merely about food (George, 2008). Although Slow Food set out with the idea of preserving ethnic and local cuisine and introducing natural and traditional production, it gradually turned into a slow movement and reached a point beyond its scope. The Slow Movement reflects the sensitivity of millions of people to the fear of exploitation by global businesses under market economy conditions and questions companies about the way they do business at the national and global levels (Bommel & Spicer, 2015). In addition to its questioning on global business, the Slow Movement has inspired societies in areas such as preserving cultural heritage, improving quality of life, achieving sustainable development and building new urbanism models. As an alternative development model, the basic assumptions of the Slow Food movement are based on local sustainability and the interdependence and balance between environment–economy–equality (Honore, 2004; Mayer & Knox, 2006).

4 Slow Tourism: A Clue for Sustainable Development

Despite the conceptual turmoil in the field of tourism, tourism researchers state that international tourism takes place in two ways. The first is mass tourism, which usually takes place in coastal areas and is commercial and seasonal. This tourism approach, based on high-density, standardised accommodation, uniform products and experience, involves many tourists who are generally strict about their cultural norms. The second approach comprises the alternative tourism forms which started to grow after the 1990s. These forms include numerous planning strategies (such as ecological, cultural responsibility or sustainability) to prevent ecological, economic and sociocultural problems that countries or destinations have to deal with due to mass tourism (Kingsbury, 2006).

Slow tourism is based on the approach of Slow Food, which emerged in Italy in 1986. The notion of slow tourism has appeared as a response by the tourism industry to problems of globalisation (Dickinson, 2009). Unlike mass tourism, slow tourism is related to integration with the destination, the ability of the tourist to establish local tour plans and the promotion of local cultures (Molz, 2009). According to Conway and Timms (2010), slow tourism, contrary to the uniform and commercial structure of mass tourism, is a form of tourism based on sustainability, real relations with local people, small-scale travel mediums and local values. In this sense, slow tourism aims to make the local communities real beneficiaries of the benefits that tourism creates by taking them into the centre in tourism decisions and investments. The loss of functionality of tourism policies, which focus solely on economic and sectoral needs, has made it imperative to pay more attention to local communities and their physical and sociocultural environments in tourism planning.

In slow tourism, travel is a pleasure and a passion, rather than a means to get from point A to point B. Slow tourism involves respecting, adapting and preserving ecological and social differences. At this point, the hospitality policy of slow tourism – within the framework of sustainability – includes social and environmental integration, increasing the attractiveness of local values and expanding the philosophy of slowness (Sezgin & Özdemir, 2016). Table 6.2 shows the differences between slow tourism and mainstream tourism. It demonstrates that slow tourism is, rather than a simple type of tourism, a Slow Food based tourism approach that can lead communities to sustainable development and sustainable tourism goals (Matos, 2004). Several authors (Conway & Timms, 2010; Dickinson

Table 6.2 Main differences between mainstream and slow tourism

Mainstream tourism	Slow tourism
Speedy transit	Slower travel times
Prevailing modes of the car and airline dominate	Wider range of modes including bus and train
Immediacy	Slowness
Resource intensive	Resource reduction
Journey is a corridor	Journey is the thing
Consumption of many attractions	Localness
Maximising visits	Staying awhile
High-carbon	Low-carbon
Commoditisation	De-commoditisation
Standardised hospitality dominates	Slow food and beverages

Source: Dickinson and Lumsdon (2010: 191).

& Lumsdon, 2010) have confirmed the inclusiveness of slow tourism. Similarly, Serdane (2017) states that slow tourism is an umbrella concept for new tourism.

5 Components of Slow Tourism in the Context of Sustainable Development

The components of the slow tourism approach have been identified as accommodation, transportation, entertainment and food and beverage (F&B), supply chain, experience, and marketing and management. Table 6.3 shows the content, applications, beneficiaries and consequences of each component of slow tourism.

Accommodation is one of the basic elements of mainstream tourism and slow tourism. However, 80% of the revenue from the all-inclusive package tours, which are often used in resorts and large hotels, go to airlines, hotels and other international companies, and not to local businesses or workers. Similarly, only US$5 out of every US$100 spent on travel in mainstream tourism remains at a developing destination (UNWTO, 2014). Likewise, in Turkey, only 10% of the money spent by tourists staying in an all-inclusive system reaches the local economy (Tourism Concern, 2015). Large hotels and all-inclusive systems also limit the contact between local people and tourists (Tavares & Kozak, 2015). According to Tourism Concern's (2015) report, of almost 1750 participants who chose an all-inclusive holiday, 32.8% never went to a local restaurant and 34.47% never went outside the resort. In slow tourism, local companies and local development are priority areas. Small and medium-sized accommodation enterprises should be used for this purpose. In this way, local entrepreneurs with limited budgets can obtain a

Table 6.3 Components and consequences of slow tourism

Component	Content	Applications	Beneficiaries	Consequences
Transportation	Travel modes	Use of local (existing) transport networks; Low-carbon transportation methods; Promoting short-distance travel	Local transport cooperatives or businesses; Natural environment	Low carbon
Accommodation	Length of stay; SMEs	Promoting longer stays at destinations; Abandoning the all-inclusive system	Local communities; SMEs at the destination	Low carbon; Reduction of economic leakage
Entertainment and F&B	Activities that involve the characteristics of and engagement with the region	Local food, local restaurants and local drinks; Slow itineraries	Local producers; Tourists	Protection of traditional occupations, products and food; Real experience
Supply chain	Use of local production and supply facilities; Use of local products	Ensuring the use of local products; Supporting local production and agriculture	Natural environment; Local communities	Biological diversity; Protection of traditional occupations; Sectoral balance; Low carbon
Experience	Direct and real contact with the local community; Use of intangible products; Quality time	Local activities (vineyard visits, olive harvests, nature trips, mushroom picking); Local festivities, weddings, soldier send-offs	Local communities; Tourists	Prevention of cultural erosion; Conversion from tourist to co-resident; Authentic experience
Marketing and management	Growth or de-growth operations; Purposeful promotion; Promotion of local cooperatives in transport, accommodation and supply chain; Use of technology; Use of slow philosophy; Carrying capacity issues	Balancing the tourist flow to ensure tourists flow to destinations that have internalised the slow philosophy; Operations to attract not only foreign tourists but also domestic tourists; Supporting cooperatives; Active use of internet and social media; Supporting Slow Food and the Cittaslow network; Considering ecological, economic, social and psychological carrying capacity	Local communities; Destination; Natural environment	Year-round tourist flow; Economic sustainability; Qualified tourists; Competitive advantage; Localisation; Increasing impact; Branding; Prevention of over-tourism

Source: Author.

direct share of the tourism cake, and by abandoning the all-inclusive system, businesses outside the hotel can benefit from direct tourism revenue. The essence of slow tourism is mental and physical engagement with the local community to enjoy leisure time in the tourism process. One way to reflect this is through long-term stays. In contrast to mass tourism, participants of slow tourism stay more than a week at a particular destination (Chao, 2015). As stated by Serdane (2017: 9), 'increased length of stay at the destination is more sustainable not only economically but also environmentally'.

Transportation has an important place in slow tourism as with other tourism approaches. The carbon emissions from transportation and the impact of tourism-based transportation on climate change make this issue more critical (Dickinson *et al.*, 2010). In addition, slow tourism attributes a very different meaning to transportation, compared to that of mass tourism. In fact, at this point, it is more appropriate to use the term 'journey' instead of 'transportation' in slow tourism. In the logic of mass tourism and mass production, transportation is the cheapest and fastest way to transfer a group of products or a tourist group to a destination. However, in slow tourism, transportation (or the journey) includes not only the destination but also the route. In this sense, reducing the frequency of travel (Caffyn, 2012), focusing on travelling shorter distances (Dickinson *et al.*, 2010), choosing a linear rather than a single-location based tour (Lumsdon & McGrath, 2011), and using local transportation networks (Dickinson *et al.*, 2011) and low-carbon transportation methods (eco-vehicles/mobilities) make travel meaningful to tourists and also reduce the pressure exerted by tourism on climate change. Moreover, the use of local (existing) transportation options contributes to poverty reduction by increasing the income of local transportation cooperatives/enterprises.

Activities and F&B are the most important areas for tourists to engage with the region. Local foods and beverages are very important as the roots of slow tourism go back to the Slow Food philosophy. One of the elements that distinguishes a destination from others and makes it unique is the local food and beverages. F&B in slow tourism not only involves the tourists visiting local restaurants, but also obtaining ingredients as co-producers, preparing food according to traditional recipes or visiting agricultural areas. The use of local agricultural products and traditional production and presentation methods contributes to the region in ecological, economic and social terms (Yurtseven *et al.*, 2010). In this way, tourists can directly experience the cultural elements of the region while also contributing to the region. In terms of the activities of tourists at the destinations, contact with the local community and cultural heritage is at the forefront. However, slow tourism also includes – without commercial concerns – land- and water-based activities, outdoor activities and nature experiences, nature and heritage discovery and gastronomy (Serdane, 2017). Thereby, slow tourism makes the tourist locally embedded in the destination rather than imposing a globalised tourist culture (Dickinson & Lumsdon, 2010).

Supply chain elements in slow tourism are based locally and regionally. Slow tourism requires a holistic approach based on sustainability and locality among stakeholders (Heitmann *et al.*, 2010). Therefore, supply elements in slow tourism should prefer local production and supply. Mayer and Knox (2006)summarise a programme that could be called a slow supply chain in Hersbruck, Germany:

> The goal is to produce and market organic apple juice by using the fruit trees in the local orchards and pasturelands. The juice is produced and marketed regionally. A third initiative aimed at linking the cultural landscape with community economic development is a project that promotes the use of local produce in traditional region-specific dishes in restaurants. Twenty-nine farmers and 17 restaurants formed a group of suppliers and gastronomic producers. The farmers supply the restaurants with their seasonal products and the restaurant offers a special menu that also identifies the producer by name and location for the consumer. A plaque at the restaurants' front door indicates their participation. (Mayer & Knox, 2006: 329)

As seen in this example, the use of a regional production and supply chain positively affects the local economy, biodiversity, traditional professions and the ecological environment due to the shortening of transport distance.

Since *experience* started being used as an economic term (Pine & Gilmore, 1998), it has been frequently attributed importance in the service industry, particularly in tourism (Chang, 2018). Slow tourism offers active experience, nurtured by increasing demand for diversity, quality and emotional experience, providing the opportunity to rediscover the spiritual and cultural dimensions of travel rather than the ordinary, standardised and packaged tourism experience of mass tourism (Calzati & Salvo, 2018). Tourists have the opportunity to directly experience parts of local community life with slow guides operating within the framework of social, cultural and economic sustainability. Thus, in the slow tourism experience, tourists become co-residents.

Marketing and management in slow, community-led tourism development encourage the change of mentality and philosophy and the re-evaluation of the changes brought about by mainstream tourism (Heitmann *et al.*, 2010). In slow tourism, optimisation is preferable to maximisation. Therefore, considering the economic, social and psychological carrying capacity of the destination, growth/marketing or de-growth/de-marketing operations are performed. Donaldson (2018) indicates that the use of 'slow' in branding and marketing contributes to success. Active use of the internet and social media rather than intermediaries should be preferred in marketing studies (Conway & Timms, 2010). In addition, purposeful promotion will bring both domestic and foreign tourists who have internalised the slow philosophy of the region. This will provide an opportunity to serve conscious tourists looking for local elements in the transport, accommodation and supply chain.

6 Conclusion

The existing business philosophy has become a threat to nature, living things, humans and even the whole world. The mass production mechanisms that emerged after the Industrial Revolution put 'speed' at the centre of production costs and exploited the earth's resources without concern. In so doing, in addition to physical resources and ecology, social structures, values, cultures and traditions were consumed. Global products, brands, production and management formed a uniform policy by imposing the same product and even the same culture on more people. Likewise, the mass tourism approach focuses on production and sales and has resulted in the uncontrolled expansion of the tourism market with uniform and low-priced products.

The overgrowth of tourism that does not take into account the carrying capacity of a destination in addition to a profit-oriented approach have resulted in various forms of damage in those areas. The growing debates on the potential and emerging negative ecological, economic and social consequences of the tourism industry have led to the necessity for an alternative approach (Tao *et al.*, 2004). In mass tourism, the benefits of tourism development are shared by a limited group, such as entrepreneurs, foreign investors and managers of tourism businesses. It creates only low-wage jobs for members of the local community in tourism businesses and ultimately increases income inequality in society. This situation results in a *high impact with low value* in tourism regions, increasing income inequality and raising living costs, which over time causes poverty problems (Lemma, 2014). Unlike mass tourism, slow tourism focuses on the local. Therefore, slow tourism refers to planned activities that are at the confluence of local entrepreneurs, SMEs, local resources, gastronomy, agriculture and local community members. The symbiotic relationship between all components results in a fair sharing of the value created by tourism. In other words, slow tourism transforms the *high effect with low value* of the dominant system into *low effect with high value*. In this context, the slowing pace of tourism has a special meaning as it is not only a way to defend and protect resources, but it also creates new opportunities for communities in socially, economically and ecologically fragile environments in both rural and urban areas (Di-Clemente *et al.*, 2014). Slow tourism, as the antithesis or antidote to exiting the tourism system, means, from a sustainability perspective, enjoying the journey itself and integrating with local culture and practices (Clancy, 2014).

Slow tourism is more attractive for communities living in rural areas with low incomes, poor working conditions, high unemployment and widespread seasonal jobs. The aim is to help maintain regular income distribution throughout the year as well as to distribute hidden unemployment and seasonal jobs over 12 months. Consequently, slow tourism has great potential to achieve ecological, economic and sociocultural sustainability through transportation, accommodation, activities and F&B, supply chain, tourist experience, and marketing and management practices. In this sense, slow tourism is a realistic option for sustainable development and poverty reduction by effectively using the tangible and intangible resources of the destinations and communities.

References

Abbasov, F., Earl, T., Jeanne, N., Hemmings, B., Gilliam, L. and Calvo Ambel, C. (2019) *One Corporation to Pollute Them All*. Transport and Environment Report. See https://www.transportenvironment.org/sites/te/files/publications/One%20Corporation%20to%20Pollute%20Them%20All_English.pdf (accessed 1 October 2019).

Alam, M.S. and Paramati, S.R. (2016) The impact of tourism on income inequality in developing economies: Does Kuznets curve hypothesis exist? *Annals of Tourism Research* 61, 111–126.

Archer, B., Cooper, C. and Ruhanen, L. (2005) The positive and negative impacts of tourism. In W.F. Theobald (ed.) *Global Tourism* (pp. 79–103). Burlington: Elsevier.

Arıkan, İ., Ünsever, I. and Halioui, S. (2016) Importance of tourism paradox, tourism equinox and tourism detox for urban environments. *Acta Economica Et Turistica* 2 (2), 221–229.

Becken, S., Frampton, C. and Simmons, D. (2001) Energy consumption patterns in the accommodation sector – the New Zealand case. *Ecological Economics* 39, 371–386.

Bianchi, R. (2017) The political economy of mass tourism and its contradictions. In D. Harrison and R. Sharpley (eds) *Mass Tourism in a Small World* (pp. 40–52). Croydon: CABI.

Bommel, V.K. and Spicer, A. (2015) Slow food as a social movement. In N.J. Smelser and P.B. Baltes (eds) *International Encyclopedia of the Social and Behavioral Sciences* (pp. 94–99). Oxford: Elsevier Science.

Britton, S.G. (1982) The political economy of tourism in the Third World. *Annals of Tourism Research* 9 (3), 331–358.

Brohman, J. (1996) New directions in tourism for Third World development. *Annals of Tourism Research* 23 (1), 48–70.

Bruns-Smith, A., Choy, V., Chong, H. and Verma, R. (2015) Environmental sustainability in the hospitality industry: Best practices, guest participation, and customer satisfaction. *Cornell Hospitality Report* 15 (3), 6–16.

Butcher, J. (2017) The morality of mass tourism. In D. Harrison and R. Sharpley (eds) *Mass Tourism in a Small World* (pp. 28–39). Croydon: CABI.

Caffyn, A. (2012) Advocating and implementing slow tourism. *Tourism Recreation Research* 37 (1), 77–80.

Calzati, V. and Salvo, P.D. (2018) Slow tourism: A theoretical framework. In M. Clancy (ed.) *Slow Tourism, Food and Cities: Pace and the Search for the 'Good Life'* (pp. 33–48). New York: Routledge.

Capocchi, A., Vallone, C., Pierotti, M. and Amaduzzi, A. (2019) Overtourism: A literature review to assess implications and future perspectives. *Sustainability* 11 (12), 3303.

Cecile, A. (2006) *Slow is Beautiful: New Visions of Community, Leisure and Joie de Vivre*. Gabriola, BC: New Society Publishers.

Chang, S. (2018) Experience economy in hospitality and tourism: Gain and loss values for service and experience. *Tourism Management* 64, 55–63.

Chao, R.F. (2015) Development of slow tourism challenge and operation architecture: A case study on Green Island, Taiwan. *Acta Oeconomica* 65 (2), 351–367.

Clancy, M. (2014) Slow tourism: Ethics, aesthetics and consumptive values. In C. Weeden and C. Boluk (eds) *Managing Ethical Consumption in Tourism* (pp. 56–69). Abingdon: Routledge.

Conway, D. and Timms, B.F. (2010) Re-branding alternative tourism in the Caribbean: The case for slow tourism. *Tourism and Hospitality Research* 10 (4), 329–344.

Dickinson, J. (2009) Adapting tourism for a lower carbon future: A slow travel approach. In M. Landre (ed.) *Transport and Tourism: Challenges, Issues and Conflicts. Proceedings of the Travel and Tourism Research Association Europe* (pp. 32–45). Breda: Centre for Sustainable Tourism and Transport.

Dickinson, J. and Lumsdon, L. (2010) *Slow Travel and Tourism*. London: Earthscan.

Dickinson, J.E., Robbins, D. and Lumsdon, L. (2010) Holiday travel discourses and climate change. *Journal of Transport Geography* 18 (3), 482–489.

Dickinson, J.E., Lumsdon, L.M. and Robbins, D. (2011) Slow travel: Issues for tourism and climate change. *Journal of Sustainable Tourism* 19 (3), 281–300.

Di-Clemente, E., Hernández-Mogollón, J.M., De Salvo, P. and Campón-Cerro, A.M. (2014) Slow tourism: An alternative model for local and tourist development. In J. Mondéjar-Jiménez, G. Ferrari, M.D.V. Segarra-Oña and A. Peiró-Signes (eds) *Sustainable Tourism: A Collection of Tools and Best Practices* (pp. 23–36). Oxford: Chartridge Books.

Donaldson, R. (2018) *Small Town Tourism in South Africa*. Cham: Springer.

Edgell, D.L. (2016) *Managing Sustainable Tourism: A Legacy for the Future* (2nd edn). New York: Routledge.

Ellis-Petersen, H. (2018) Visitors flocked to Maya Bay on Ko Phi Phi Leh island after it was in Leonardo DiCaprio film. *The Guardian*, 3 October. See https://www.theguardian.com/world/2018/oct/03/thailand-bay-made-famous-by-the-beach-closed-indefinitely (accessed 12 November 2018).

European University Centre for Cultural Heritage (2018) *Manila Declaration on World Tourism*. See https://www.univeur.org/cuebc/downloads/PDF%20carte/65.%20Manila.PDF (accessed 15 April 2018).

Filimonau, V., Dickinson, J. and Robbins, D. (2014) The carbon impact of short-haul tourism: A case study of UK travel to Southern France using life cycle analysis. *Journal of Cleaner Production* 64, 628–638.

George, A. (2008) *The Slow Food Story: Politics and Pleasure*. London: Pluto Press.

Ghaderi, Z. (2013) Sustainable tourism development planning principles in rural areas. In *4th International Conference of Economic Sciences; Quality of Life, Sustainability and Locality, Kaposvár University, Kaposvár, Hungary, 9–10 May*.

Gössling, S., Peeters, P., Hall, C.M., Ceron, J.P., Dubois, G., Lehmann, L.V. and Scott, D. (2012) Tourism and water use: Supply, demand, and security. An international review. *Tourism Management* 33 (1), 1–15.

Hall, D. and Brown, F. (2006) *Tourism and Welfare: Ethics, Responsibility and Sustained Well-being*. Wallingford: CABI.

Heinonen, S., Halonen, M. and Daldoss, L. (2006) Slow housing – competitive edge for innovative living environments. *Fennia* 184 (1), 91–104.

Heitmann, S., Robinson, P. and Povey, G. (2010) Slow food, slow cities and slow tourism. In P. Robinson, S. Heitmann and P. Dieke (eds) *Research Themes in Tourism* (pp. 114–127). Wallingford: CABI.

Holden, A. (2008) *Environment and Tourism* (2nd edn). New York: Routledge.

Honore, C. (2004) *In Praise of Slowness: How a Worldwide Movement is Challenging the Cult of Speed*. San Francisco, CA: Harper.

Howard, C. (2012) Speeding up and slowing down: Pilgrimage and slow travel through time. In S. Fullagar, K. Markwell and E. Wilson (eds) *Slow Tourism: Experiences and Mobilities* (pp. 11–24). Bristol: Channel View Publications.

Incera, A.C. and Fernández, M.F. (2015) Tourism and income distribution: Evidence from a developed regional economy. *Tourism Management* 48, 11–20.

Kaya, O. (2016) Turizm Sistemi ve Birey. In E.D.C. Aslan (ed.) *Turizmin Sosyal Psikolojik Dinamikleri* (pp. 324–340). Çanakkale: Paradigma Akademi.

Kingsbury, P. (2006) Corporate environmental sustainability: Sandals Resorts International in Jamaica. In J. Pugh and J.J. Momsen (eds) *Environmental Planning in the Caribbean* (pp. 111–128). Farnham: Ashgate.

Knox, P.L. (2005) Creating ordinary places: Slow cities in a fast world. *Journal of Urban Design* 10 (1), 1–11.

Kotios, A., Plageras, P., Galanos, G., Koutoulakis, M. and Saratsis, Y. (2013) Water and tourism. In A. Holden and D. Fennell (eds) *The Routledge Handbook of Tourism and the Environment* (pp. 435–446). Abingdon: Routledge.

Kundera, M. (1997) *Slowness* (L. Asher, trans.). New York: Harper Collins.

Lea, J. (2001) *Tourism and Development in the Third World*. London: Routledge.

Lee, S. (2009) Income inequality in tourism services-dependent counties. *Current Issues in Tourism* 12 (1), 33–45.

Lemma, A.F. (2014) *Tourism for Poverty Reduction in South Asia: What Works and Where Are the Gaps?* London: DFID.

Lenzen, M., Sun, Y.Y., Faturay, F., Ting, Y.P., Geschke, A. and Malik, A. (2018) The carbon footprint of global tourism. *Nature Climate Change* 8 (6), 522–528.

Lipschutz, D.R. (2002) The clash of governmentalities: The fall of the UN republic and America's reach for imperium. *Contemporary Security Policy* 22 (2), 214–231.

Lumsdon, M.L. and McGrath, P. (2011) Developing a conceptual framework for slow travel: A grounded theory approach. *Journal of Sustainable Tourism* 19 (3), 265–279.

Matos, R. (2004) Can 'slow tourism' bring new life to Alpine regions? In K. Weimar and C. Mathies (eds) *The Tourism and Leisure Industry: Shaping the Future* (pp. 93–104). New York: Routledge.

Mayer, H. and Knox, P.L. (2006) Slow cities: Sustainable places in a fast world. *Journal of Urban Affairs* 28 (4), 321–334.

Molz, J.G. (2009) Representing pace in tourism mobilities: Staycations, slow travel and the amazing race. *Journal of Tourism and Cultural Change* 7 (4), 270–286.

Oh, H., Assaf, A.G. and Baloglu, S. (2014) Motivations and goals of slow tourism. *Journal of Travel Research* 55 (2), 205–219.

Paul, B.D. (2012) The impacts of tourism on society. *Annals of Faculty of Economics, University of Oradea, Faculty of Economics* 1 (1), 500–506.

Pine, B.J. and Gilmore, J.H. (1998) Welcome to the experience economy. *Harvard Business Review*, July–August. See https://hbr.org/1998/07/welcome-to-the-experience-economy (accessed 15 June 2017).

Postma, A. (2013) When the tourists flew in: Critical encounters in the development of tourism. Unpublished PhD thesis, Faculty of Spatial Sciences, University of Groningen.

Ramukumba, T. (2015) The contribution of guest houses to economic growth and employment as key components of local economic development in the Eden District area. Unpublished PhD thesis, Nelson Mandela Metropolitan University.

Rettinger, R. and Apollo, M. (2017) Enriching tour-operators' offer to go beyond sea, sun and sand: The case of the Dominican Republic. In R. Efe and M. Öztürk (eds) *Contemporary Studies in Environment and Tourism* (pp. 60–81). Newcastle: Cambridge Scholars.

Scheyvens, R. and Hughes, E. (2019) Can tourism help to 'end poverty in all its forms everywhere'? The challenge of tourism addressing SDG1. *Journal of Sustainable Tourism* 27 (7), 1061–1079.

Schouten, M. (2003) *Development in the Drought: The Incompatibility of the Ebro Water Transfer with Sustainable Development in the Southeast Region of Spain*. Madrid: World Wildlife Fund (WWF)/Adena. See http://wwf.panda.org/wwf_news/?7048/

Development-in-the-drought-The-incompatibility-of-the-Ebro-water-transfer-with-sustainable-development-in-the-Southeast-region-of-Spain (accessed 1 October 2019).

Serdane, Z. (2017) Slow tourism in slow countries: The case of Latvia. Unpublished PhD thesis, Salford Business School, University of Salford.

Sezgin, M. and Özdemir, S. (2016) Yavaş Turizm. In E. Akgöz (ed.) *Turizm Paradigmaları* (pp. 1–24). Ankara: Detay Yayıncılık.

Singer, P. (1993) *Practical Ethics* (2nd edn). Cambridge: Cambridge University.

Slow Food (2019) *Our History*. See https://www.slowfood.com/about-us/our-history/ (accessed 10 April 2019).

Spenceley, A. (2005) Nature-based tourism and environmental sustainability in South Africa. *Journal of Sustainable Tourism* 13 (2), 136–170.

Tao, C.H., Eagles, P.F. and Smith, S.L.J. (2004) Implications of alternative definitions of ecotourists. *Tourism Analysis* 9 (1–2), 1–13.

Tavares, J.M. and Kozak, M. (2015) Tourists' preferences for the all-inclusive system and its impacts on the local economy. *European Journal of Tourism, Hospitality and Recreation* 6 (2), 7–23.

Toffler, A. (1980) *Third Wave*. New York: Bantam Books.

Tourism Concern (2015) *The Perceived Impacts of All-inclusive Package Holidays on Host Destinations: A Consumer Survey*. See https://tourismconcern.org.uk/wp-content/uploads/2015/02/The-perceived-impacts-of-all-inclusives-%E2%80%93-survey-results.pdf (accessed 12 February 2018).

UNWTO (2014) *AM Reports, Volume 9 – Global Report on Adventure Tourism*. Madrid: World Tourism Organization.

UNWTO (2018a) *'Overtourism'? Understanding and Managing Urban Tourism Growth beyond Perceptions*. Madrid: United Nations World Tourism Organisation. See https://www.e-unwto.org/doi/pdf/10.18111/9789284420070 (accessed 1 March 2019).

UNWTO (2018b) *UNWTO Tourism Highlights 2018 Edition*. Madrid: United Nations World Tourism Organisation. See https://www.e-unwto.org/doi/pdf/10.18111/9789284419876 (accessed 1 April 2019).

UNWTO (2019) *Key Tourism Figures*. Madrid: United Nations World Tourism Organisation. See http://cf.cdn.unwto.org/sites/all/files/pdf/unwto_key_figures_barom_may2019_en.pdf (accessed 2 June 2019).

Urry, J. (2002) *The Tourist Gaze*. London: Sage.

Wen, J. and Tisdell, C. (1997) Regional inequality and tourism distribution in China. *Pacific Tourism Review* 1 (2), 119–128.

Yasumura, K. (1994) Sociological sphere of tourism as a social phenomenon. Revised version of a draft for the Workshop on Tourism Studies of the Department of Tourism, Hokkai Gakuen University of Kitami. See https://www.academia.edu/2262641/The_Sociological_Sphere_of_Tourism_as_a_Social_Phenomenon (accessed 11 December 2015).

Yurtseven, H.R., Kaya, O. and Harman, S. (2010) *Yavaş Hareketi* (*Slow Movement*). Ankara: Detay Yayıncılık.

Zhuang, X., Yao, Y. and Li, J. (2019) Sociocultural impacts of tourism on residents of World Cultural Heritage Sites in China. *Sustainability* 11 (3), 840.

7 Agritourism and Ecotourism as an Innovative Direction for Sustainable Rural Development

Agnieszka Gawlik and Barbara Woś

The most valuable elements of rural capital, aside from social factors, are the natural environment and landscape. A village provides an alternative to the city: a place of rest and preservation of culture and tradition, and a source of ecological safety. The preservation and maintenance of the ecological and scenic features of rural areas is becoming a strategic objective in increasing their appeal for tourism. Agritourism and ecotourism in areas with an attractive physical and geographical environment are becoming leading elements of multifunctional, sustainable development of rural areas. Increasingly, on-farm tourism activities are shifting from side to core activities. The basic condition for this development is an attractive landscape with, in particular, the presence of forests and water reservoirs (both natural and artificial). As the expectations of potential customers rise, other increasingly important factors in the development of tourism in rural areas are the standard of accommodation, meal provision and commercial and technical infrastructure (media). Hence, intensive development of accommodation in rural areas is supported by local government organisations, seeing it as an opportunity to activate these areas. Many projects financed from national funds and the EU budget serve to increase the attractiveness of tourism, as well as to develop multifunctional tourism and recreation infrastructure in rural areas (Drobek et al.*, 2016). Support for tourist activity is also indicated as one of several measures that can be applied in programmes backing rural development from the European Agricultural Fund.*

1 Introduction

The concept of sustainable development was created in the 1980s as a response to noticeable environmental degradation and an increase in disproportions in economic development and access to basic environmental resources. A widely quoted definition of sustainable development appeared for the first time in the UN report *Our Common Future* (WCED, 1987). It defined the concept as development that ensures that the needs of present generations are met without negatively impacting the ability to meet the needs of the next generations. According to this definition, economic development should be based on rational and effective use of available resources and technologies while limiting as much as possible the adverse impact of humans on the natural environment. Such development should provide the opportunity to meet the social, economic and ecological needs of contemporary and future generations (Collins, 1999; Kowalczyk, 2011).

Tourism is a field of the economy in which the implementation of sustainable development is crucial. Tourist activity is based on the use of the potential of the natural and cultural environment, creating conditions for the development of various forms of tourism. The dynamic development of tourism, mass trips to the most attractive regions of the world, excessive development of tourist infrastructure and increasing environmental pollution have led to the degradation of many attractive tourist areas. The implementation of sustainable development in tourism has become an absolute necessity.

According to the World Tourism Organization, sustainable tourism should take full account of its current and future economic, social and environmental effects, while also taking into account the needs of tourists, the tourism industry, the natural environment and the host community (UNEP & UNWTO, 2005). It should respect the natural environment, ensure the long-term preservation of natural and cultural values and enable the economic development of local communities. To achieve this, it is necessary to develop forms of tourism based on respect for the natural diversity of the visited areas, as well as the cultural identity of local communities. Sustainable tourism development should be treated as a way to manage the tourist economy in order to find a compromise between the needs of local communities and economic entities operating in the field of tourism and the requirements of environmental protection (Królikowska, 2017; Xing & Dangerfield, 2011).

The assumption is that in sustainable tourism the local community participates in the creation of a tourist offer, from the planning stage through to the end stage, with a fair distribution of profits, a large part of which should go to the local population. Adverse impacts on the environment caused by tourist activity should be minimised through waste reduction, energy and water saving and the elimination of substances that are dangerous to the environment. Sustainable tourism should also shape pro-ecological behaviour among service providers, clients and the local community (Dolnicar *et al.*, 2019; Zaremba, 2008). Many forms of tourism follow this path. Some of the best examples are agritourism and ecotourism. These forms represent 'the desire for tourism to be not only a positive, dynamic development factor but also a viable solution to keep the environment untainted' (Dorobanţu & Nistoreanu, 2012). In addition, they stimulate the development of other activities, such as crafting, local food production or agriculture, leading to increased revenue in the corresponding areas (Ana, 2017).

Support of sustainable agritourism and ecotourism facilitates the fight against poverty caused by high unemployment rates, the decline in the profitability of agricultural production on small family farms and broadly understood social exclusion (Bricker *et al.*, 2013). Through the development of tourism, new jobs are created (small enterprises, service facilities, handicraft workshops), which offer the potential for additional earning and a gradual exit from poverty. In many areas affected by poverty, pro-poor tourism strategies have been implemented to support the local development of tourism aimed at the professional activation of excluded groups, including women, the poorly educated and the elderly (Suntikul *et al.*, 2009). The opportunity to earn income improves self-esteem and increases the chance of a better future. Micro-entrepreneurship supported by the state, social organisations and local tourism agendas affects local development and the protection of the cultural and natural heritage of the region, which is the basis for creating a tourist product (Goodwin, 2002, 2007).

2 Agritourism and Ecotourism as Forms of Rural Tourism

Rural tourism is a growing sector and offers attractive opportunities that arise from the emerging trends in tourism demand, which tend to favour less massified forms and pay more attention to the values of nature, culture, food and the countryside. Rural tourism incorporates the tourist services in rural areas – services involving investors, tour operators and local and central governments. These services include accommodation, meals (with a focus on traditional local cuisine) and all leisure activities that meet the desires of tourists (Barbu, 2013), who can usually be deemed nature-lovers. Rural tourism includes all forms of tourism that take place in rural areas or rural communities practised in small family-owned establishments. It is seen as an alternative to or as complementary to mass tourism. It highlights natural areas, local practices, culture and gastronomy.

Agritourism and ecotourism can be valuable mechanisms for alleviating poverty and exclusion in rural areas in both developed and developing countries. These branches of tourism allow the full use of the potential of the natural environment, economic development of rural areas (Rogerson &

Rogerson, 2014), creation of new jobs and the professional activation of women. The development of direct sales, including farm products and services, generates revenue directly to the owners of farms and their employees, bypassing numerous intermediaries, which significantly improves the material situation of farms and the entities cooperating with them. The development of tourism in rural areas enables the creation of new jobs, both year-round and seasonal. The scope of services that include guest care is, in many cultures, the domain of women, who may find employment and opportunities for self-fulfilment in the development of these branches of tourism. Creating an ecotourism or agritourism product, as opposed to a mass tourism product, requires significantly fewer financial resources, meaning that it can also be available to the poor and can be the basis for cooperation across several farms or even the entire village (Lyttleton & Allcock, 2002).

Agritourism and ecotourism are the two main forms of rural tourism that particularly bring benefits both for tourists and for the local community:

- Agritourism, in which tourists see and participate in traditional agricultural practices without destroying the ecosystem, the host base.
- Ecotourism, a form of tourism that features the natural resources while maintaining the values and welfare of the local population.

Agritourism refers to an activity conducted in a rural area, based at a single farm or group of farms, providing accommodation, meals, educational and recreational services and allowing tourists to actively participate in activities carried out at the site (e.g. production activities on the farm and also processing of agricultural products in the household).

Ecotourism is a form of active tourism that incorporates both outings for physical activities (hiking, horse riding, canoe trips, ski trips) and also intellectual activities related to learning about the nature and traditions of the region, on a deeper level than the information available in guides and on tourist maps. The goals of ecotourism are to allow tourists to get closer to nature, to understand the relationship between humans and the environment, to understand the laws and dependencies that govern the natural environment and to learn how to function properly in it (Dorobanțu & Nistoreanu, 2012; Kulczyk & Lewandowski, 2004). Additionally, it aims at enabling knowledge and admiration of areas of natural value, preserving the harmony of natural ecosystems and the cultural identity of local communities, providing financial resources to protect the value of natural and cultural heritage, and delivering real economic and social benefits to local communities (Ana, 2017; Zaremba, 2009). Ecotourism is more related to the specific activities carried out in a given area, as opposed to agritourism which is associated with a specific place – a farm, for instance. The combination of both forms of tourism gives new opportunities for the development of an interesting tourist offering that brings together the local natural and scenic attributes and the existing accommodation and catering facilities.

Ecotourists need accommodation and catering facilities. The introduction of ecotourism elements increases the attractiveness of the offerings of agritourism farms, enriching them with cognitive aspects and active activities (nature tours, herbal workshops, bird watching). Both forms of tourism, when well planned and carried out, complement each other, allowing full use of the natural environment and human resources in creating an attractive tourist product.

The majority of agritourists are people seeking peace and quiet away from the hustle and bustle of big cities and crowded hotels. Others are looking for safe leisure activities and access to healthy food and recreation space. A few are low income earners who are looking for comparatively cheap accommodation and leisure options. Agritourists are often interested in farming, crafting, folklore, natural agricultural products and gastronomy.

Ecotourists are usually educated people with clearly defined destinations in mind, expecting specific elements of the tourist offer and focused on active recreation in which the cognitive function plays a significant role. These are tourists who are willing to pay more for properly organised, attractive leisure, often consciously choosing places that have a lower standard of services but that enable the visitor to learn about the specifics of nature and culture of the region. They expect meaningful community participation, slow travel, high-quality experiences, picturesque natural elements, gastronomy, traditions and routes that allow them to feel as if they are locals (Ana, 2017).

3 The Impact of Agritourism and Ecotourism on the Development of Rural Areas

Agritourism develops in rural areas, and often at the periphery, where the basic assets are a natural environment altered only minimally by humans, small, peaceful and quiet villages, and cultural values typical of the specific geographical region and local communities. Recreation in the countryside allows respite from big crowded cities, access to natural, unprocessed food and regional cuisine, the opportunity to observe the work on the farm, and contact with animals. The rural environment can also be the basis for ecotourism nature expeditions and wildlife observations. Such a wide range of attractions related directly to nature is an interesting alternative to mass tourism.

The development of agritourism supports the achievement of the economic and social potential of the village. It makes it possible to move away from the monofunctional model of agriculture towards multifunctional rural development (Sammel, 2016). The possibilities of such development are extremely important in light of the economic problems in many countries associated with the falling profitability of agricultural production, high unemployment and the emigration of young people to cities. The development of agritourism provides an opportunity for socioeconomic recovery in rural areas through the creation of new jobs, the possibility of obtaining income from non-agricultural activities and the use of natural and cultural values for the creation of a tourist product (Ciolac *et al.*, 2013; Roszkowska-Mądra & Roman, 2015).

The basis of the tourist product is the richness and uniqueness of the natural environment (landscape diversity, the presence of natural watercourses and reservoirs with recreational uses, peace and quiet, clean air, presence of forest). This makes it a priority, for both owners of agritourism farms and their guests, to support the preservation of the natural environment.

The use of the human potential of the village is also important. The creation of a new range of services directly related to the tourism industry (tourist equipment rentals, catering facilities and organisation of occupational workshops) provides opportunities for employment of unemployed people, pensioners or people wishing to work for a shorter period of time. It also creates an opportunity to use the talents of representatives of the local community (teachers, artisans, folk artists), and thereby increase self-esteem.

The development of an agritourism offer also provides the possibility for creative action aimed at starting a new range of services (spa services, cosmetic workshops, wickerwork workshops – Figures 7.1 and 7.2) that may encourage young people to stay in the countryside or attract people who want to use their skills in creating an innovative offer of agri- and ecotourism. In an era when people wish to return to nature and seek new experiences, leisure at agritourism farms is becoming more and more popular.

Figure 7.1 Workshop for decorating wooden toys: Agritourism farm Organistówka in Leszczawa Dolna, Poland

Photograph: B. Woś.

Figure 7.2 Herbal workshop: Agritourism Uroczysko in Dylaki, Poland
Photograph: B. Woś.

Growing competition and constantly changing tourist preferences mean that the quality of tourist services has become more and more important. The contemporary tourist market is a typical consumer market. Success in market competition requires offering a product that satisfies the needs of tourists and meets their expectations (Ciolac *et al.*, 2019; Gawlik & Woś, 2015). The offer of a holiday in the countryside includes not only the accommodation or the opportunity to participate in the daily life of the farm but also the opportunity to pursue a passion, such as horse riding, fishing, mushroom picking or active leisure. Competition may concern the standard and the range of services, the visual attractiveness of the farm or the uniqueness of the offer (Ciolac *et al.*, 2019; Kachniewska, 2002; Woś *et al.*, 2018). This often leads to the search for a market niche and, as a result, to the specialisation of farms (e.g. farms with wine/honey production and tasting, herbal or vegan farms, farms with extensive recreation activities for children or those that accommodate only adults). Specialisation is used where there are many agritourism farms with a similar standard of services. Often, agritourism farms cooperate by creating shared packages of services available to guests, which allow the extension of the offer without the need for additional investments.

The basis for the creation of such an offer is the full use of the features of the natural environment (e.g. geological workshops in the Kaczawskie Mountains in Poland, rich in semi-precious stone deposits; herbal workshops in the Podkarpackie region in Poland, characterised by rich biodiversity and the presence of unique plant communities – see Figure 7.2) as well as the knowledge and competence of the hosts in the cultural specifics of the region (preparation of regional dishes) and disappearing skills and crafts (wickerwork, herbalism, blacksmithing). Organic food and regional and traditional products are also promoted, often produced on the farm and constituting one of the elements of its tourist offer (goat cheeses, locally pressed fruit juices, tinctures, fruit confitures). More and more agritourism farms sell package offers including accommodation, meals and attractions, paying particular attention to the quality of services provided.

The development of agritourism and ecotourism can also be a significant way to improve the economic situation and rebuild tourist reception areas damaged by natural disasters. An example is the Belmont Estate Fair trade-certified agritourism business in Grenada (the Caribbean). Ecotourism activities have been revitalised in the region, which was damaged by Hurricane Ivan in 2004. Belmont Estate offers a unique tour of cocoa plantations which allows visitors to experience chocolate production – from the field to fermentation to the final product. Visitors can take part in cultural events, animal petting, tours of chocolate-making facilities, observation of cocoa at various stages of processing, and chocolate and cocoa tea sampling. Belmont Estate participates in the Grenada Goat Dairy Project, which aims to help farmers earn income from supplying high-quality goat milk, goat cheese, yogurt and ice cream and selling these products to visitors, local markets and restaurants (Rhiney *et al.*, 2015).

Another project that aims to improve the quality of life of the local community is the Nam Ha Ecotourism Project, started in 1999 in the town of Luang Namtha in Luang Namtha Province (Laos). Luang Namtha Provincial Tourism Office offers several trips to Akha, Tai Daeng, Khamu and Lahu villages. Trips include overnight stays in the villages, one-day forest walks, more challenging two- or three-day trails with local guides and one-day boat trips. Village residents offer tourists accommodation, meals and guide services, and earn money from selling handicrafts and other services, such as massages (Harrison & Schipani, 2007).

An interesting initiative existing in the Polish agritourism market is the Nationwide Network of Educational Farms. The network is used to disseminate the idea of education in an agricultural holding and to promote the activities of educational farms. The farms associated with this network organise and conduct workshops designed for children and adolescents. The topics of the workshops are related to the profile of the farm (plant and animal production or processing of agricultural products), ecological and consumer awareness, the heritage of the material culture of the village, traditional occupations, handicrafts and folk art. The participants of the network benefit from the promotion of their offers online as well as advisory and training support. They also have the opportunity for mutual exchange of experiences. The network enables interesting educational campaigns, permanent cooperation with schools and kindergartens, and additional income outside the tourist season.

Another form of agritourism farm activity, providing accommodation, meals and recreation facilities, may be social tourism. Care farms combine agricultural and agritourism activities with care for people with disabilities, the elderly or those in need of support. In addition to typical agricultural activities, therapeutic, care and integration activities may be conducted on the care farm. Being in nature, participating in farm work and awareness of one's role as a useful member of a group are factors that play a huge role in ensuring well-being and improving health regardless of age. As part of their social functions, care farms may fulfil the following tasks: re-education and rehabilitation, therapy, professional activation, pedagogical support and care services. Such activities may be conducted through a day stay, multi-day therapeutic stay or permanent 24-hour care. This form of action perfectly fits into the idea of sustainable development of rural areas. It allows for optimal use of the farm's potential and integration of rural and urban communities. It is also a response to market demand as it presents a care offer dedicated to an ageing society. Care activities carried out on a rural farm are an interesting alternative to day care homes and standard care centres. They also allow for professional activation of the rural community and creation of new jobs. Due to the high social importance of these setups, there are many possibilities for co-financing from government programmes (e.g. Multiannual Government Program WIGOR, Government Program of Social Activity of the Senior People 2014–2020, Operational Program Knowledge Education Development – Incubators of Social Innovation in Poland) as well as projects financed by the EU. In Europe, such farms operate in the Netherlands, Belgium, France, Norway, Germany, Italy and the UK.

4 European Union Instruments to Encourage the Development of Agritourism and Ecotourism

The European Commission in Communication ID 2010/252, 'Europe, the world's No 1 tourist destination – a new political framework for tourism in Europe', identified sustainability as a fundamental aspect of the quality and competitiveness of European tourism. In this communication, the Commission focuses on two key concepts that mutually influence each other: the need for a sustainable approach and the need to boost the competitiveness of the European tourism sector. The sector's competitiveness is closely linked to its sustainability, as the quality of tourist destinations is strongly influenced by their natural and cultural environment and their integration into a local community. Sustainability principles refer to the environmental, economic and sociocultural aspects of tourism development, and a suitable balance must be established between these three dimensions to guarantee long-term sustainability. The sustainability of tourism covers, thus, a number of different aspects: the responsible use of natural resources, careful consideration for the environmental impact of activities (production of waste, pressure on water supply, land and biodiversity, etc.), the use of 'clean' energy, protection of the heritage and preservation of the natural and cultural integrity of destinations, the quality and sustainability of jobs created, and local economic fallout or customer care.

The instrument for the implementation of the sustainable development of rural areas, and consequently sustainable tourism, is the Rural Development 2014–2020 policy, worth €100 billion and funded through the European Agricultural Fund for Rural Development (EAFRD; European Commission, 2017b). There are 118 different rural development programmes (RDP) implemented in the 28 Member States for this period (20 single national programmes and eight Member States having two or more regional programmes). They focus on agricultural sector competitiveness, sustainable management of natural resources and climate action as well as balanced territorial development of the European rural areas to improve the quality of life in these regions by fostering the diversification of economic activities. EU countries implement their rural development programmes based on the needs of their territories. Among other things, EAFRD funds finance the establishment of businesses active within rural tourism, the development and promotion of agritourism, and capitalisation on the cultural and natural heritage of rural regions (European Commission, 2010).

Another instrument for implementing sustainable development in rural areas is the LEADER programme, which was launched in 1991 and has proved itself to be of significant importance to rural tourism. It provides EU rural communities with the necessary resources to enable local partners to actively engage in and direct community-led local development. Among the great range of projects and activities undertaken and funded by LEADER, those that aim to implement rural tourism are the most popular (Lehmeier, 2010).

The tourism sector is influenced and supported by various sectoral policies. The EU's goals in the tourism field can be found in the EU's Directorate General Enterprise and Industry. One of its main initiatives is the European Destinations of Excellence (EDEN), which promotes sustainable tourism development models across the EU. It was launched in 2006. The initiative is based on national competitions and promotional campaigns that result in the selection and promotion of a tourist 'destination of excellence' for each participating country.

Through the selection and promotion of emerging, little-known destinations located in the 28 EU countries and candidate countries, EDEN effectively achieves the objective of drawing attention to the values, diversity and common features of European tourist destinations. The key feature of the selected destinations is their commitment to the social, cultural and environmental sustainability of tourism. The EDEN initiative helps to spread the sustainable practices used in the awarded destinations across the EU. Since 2006, 158 EDEN destinations have been elected, based on given themes and aiming at promoting sustainable tourism. The topics are chosen by the European Commission in close cooperation with national tourism authorities and are always related to sustainable tourism development, whether from a cultural, economic, environmental or local involvement point of view. The project started with the theme, 'Best emerging rural destination of excellence' (2007), followed by 'Tourism and local intangible heritage' (2008), 'Tourism and protected areas' (2009), 'Aquatic tourism' (2010), 'Tourism and regeneration of physical sites' (2011), 'Accessible tourism' (2013), 'Tourism and local gastronomy' (2015) and 'Cultural tourism' (2017). The chosen destinations are promoted as a European brand and enable the establishment of the EDEN network.

A preparatory action launched in 2009 called Sustainable Tourism is another programme that has contributed to the development of rural forms of tourism, promoting routes, itineraries and trails throughout Europe, such as the EuroVelo 13, also known as the Iron Curtain Cycling Trail (which led to the creation of over 10,000 cross-border cycling routes), spanning over 70,000 km, with an estimated completion date for the entire network being 2020.

5 Conclusion

Most European countries, and beyond, see rural development as an opportunity to fight poverty and to ensure the sustainability of the rural environment. The main purpose of sustainable development is to guide actions in a responsible and harmonious way, taking into account all the environmental and societal consequences as well as economic purposes (Font & McCabe, 2017). Agritourism is an activity that links the economic, social and environmental elements of sustainability, strongly related to local communities and their attitudes towards tourism (Apollo & Rettinger, 2019; Muresan *et al.*, 2016). It can be seen as an innovative and diversifying strategy for farms, with many economic and non-economic benefits for farmers, visitors and communities (Bramwell, 1994; Turnock, 1999).

The implementation of the pro-poor tourism strategy through the supported development of agritourism and ecotourism in poverty-stricken areas encourages cooperation between local governments, representatives of national and foreign tourism, and social organisations, whose main goal is to build shared prosperity and reduce poverty. It creates opportunities for direct and indirect employment in one of the most labour-intensive sectors of the economy – tourism. The high demand for employees with different qualifications creates the possibility of employing poorly educated employees; it also provides an opportunity to increase their competence and develop new skills during training and internships.

It is not surprising, therefore, that in European countries agritourism (and ecotourism) have become a priority in local development policies and are supported by EU funds. Agritourism and ecotourism generate significant contributions to the process of sustainable rural development, with specific effects on the environment, agricultural heritage or economic growth (Ciolac *et al.*, 2013, 2019). These forms of tourism stop rural out-migration by: keeping farmers on the land, improving the use of rural resources; enhancing environmental conservation and management; promoting typical rural products; supporting rural traditions and cultural initiatives; and increasing the economic development of agricultural areas (creating revenue growth, access to work, exchanges between rural and urban areas, multiplier effects on direct investment, diversification of economic activities, increased value of properties in the area, a ready infrastructure and opportunities for other economic developments). They can also strengthen the local structure through working groups, develop youth and social tourism and enhance the relationship between city and countryside (Ciolac *et al.*, 2019).

References

Ana, M.I. (2017) Ecotourism, agro-tourism and rural tourism in the European Union. *Cactus Tourism Journal* 15 (2), 6–14.

Apollo, M. and Rettinger, B. (2019) Mountaineering in Cuba: Improvement of true accessibility as an opportunity for regional development of communities outside the tourism enclaves. *Current Issues in Tourism* 22 (15), 1797–1804.

Barbu, I. (2013) Approach to the concept of rural tourism. *Scientific Research. Agricultural Management* 15 (4), 125–128.

Bramwell, B. (1994) Rural tourism and sustainable rural tourism. *Journal of Sustainable Tourism* 2 (1–2), 1–6.

Bricker, K.S., Black, R. and Cottrell, S. (eds) (2013) *Sustainable Tourism and the Millennium Development Goals: Effecting Positive Change* (1st edn). Burlington, MA: Jones & Bartlett.

Ciolac, R., Iancu, T., Rujescu, C., Milin, A., Merce, I., Marin, D., Dincu, A.M. and Stanciu, S.O. (2013) Agro-tourism in European Mountain Areas. *Rural Development Proceeding* 6, 80–85.

Ciolac, R., Adamov, T., Iancu, T., Popescu, G., Lile, R., Rujescu, C. and Marin, D. (2019) Agritourism – a sustainable development factor for improving the 'health' of rural settlements. Case study Apuseni Mountains Area. *Sustainability* 11 (5), 1467.

Collins, A. (1999) Tourism development and natural capital. *Annals of Tourism Research* 26 (1), 98–109.

Dolnicar, S., Cvelbar, L.K. and Grün, B. (2019) A sharing-based approach to enticing tourists to behave more environmentally friendly. *Journal of Travel Research* 58 (2), 241–252.

Dorobanţu, M.R. and Nistoreanu, P. (2012) Rural tourism and ecotourism – the main priorities in sustainable development orientations of rural local communities in Romania. *Economy Transdisciplinary Cognition* XV (2067–5046), 259–266.
Drobek, W., Gawlik, A. and Woś, B. (2016) Obiekty hotelarskie na terenach wiejskich jako czynnik rozwoju Wsi Opolskiej. In K. Heffner and B. Klemens (eds) *Studia Komitetu Przestrzennego Zagospodarowania Kraju PAN Obszary wiejskie – wiejska przestrzeń i ludność, aktywność społeczna i przedsiębiorczość, CLXVII* (pp. 172–179). Warszawa: IGiPZ PAN and PTG.
European Commission (2010) *Europe, the World's No 1 Tourist Destination – a New Political Framework for Tourism in Europe*. Brussels: European Commission.
European Commission (2017a) *About EDEN*. See https://ec.europa.eu/growth/sectors/tourism/eden/about_en (accessed 15 May 2019).
European Commission (2017b) *Rural Development 2014–2020*. See https://ec.europa.eu/agriculture/rural-development-2014–2020_en (accessed 15 May 2019).
EuroVelo (2019) *EuroVelo*. See www.eurovelo.org/ (accessed 15 May 2019).
Font, X. and McCabe, S. (2017) Sustainability and marketing in tourism: Its contexts, paradoxes, approaches, challenges and potential. *Journal of Sustainable Tourism* 25 (7), 869–883.
Gawlik, A. and Woś, B. (2015) The use of CRM instruments in agritourism – on the example of agritourism farms in the Opole Province. *Scientific Journal of University of Szczecin Service Management* 16 (2), 49–56.
Goodwin, H. (2002) Local community involvement in tourism around national parks: Opportunities and constraints. *Current Issues in Tourism* 5 (3–4), 338–360.
Goodwin, H. (2007) Indigenous tourism and poverty reduction. In R. Butler and T. Hinch (eds) *Tourism and Indigenous Peoples* (pp. 84–94). Oxford: Butterworth-Heinemann.
Harrison, D. and Schipani, S. (2007) Lao tourism and poverty alleviation: Community-based tourism and the private sector. *Current Issues in Tourism* 10 (2–3), 194–230.
Kachniewska, M. (2002) *Zarządzanie jakością usług turystycznych*. Warsaw: Difin.
Kowalczyk, M. (2011) Wskaźniki zrównoważonego rozwoju turystyki. *Człowiek i Środowisko* 35 (3–4), 35–50.
Królikowska, K. (2017) Narzędzia wdrażania zasad turystyki zrównoważonej na obszarach górskich. *Rozprawy naukowe Akademii Wychowania Fizycznego we Wrocławiu* 56, 36–51.
Kulczyk, S. and Lewandowski, W. (2004) Ekoturystyka 'zieloną' ścieżką do Europy. In M. Kistowski (ed.) *Studia ekologiczno-krajobrazowe w programowaniu rozwoju zrównoważonego. Przegląd polskich doświadczeń u progu integracji z Unią Europejską* (pp. 115–121). Gdańsk: Uniwersytet Gdański.
Lehmeier, H. (2010) Tourism policy and regional development in the European Union. *Geographica Timisiensis* 19 (2), 121–131.
Lyttleton, C. and Allcock, A. (2002) *Tourism as a Tool for Development: UNESCO-Lao National Tourism Authority Nam Ha Ecotourism Project-External Review July 6–18 2002*. Vientiane: NTAL and UNESCO.
Muresan, I.C., Oroian, C.F., Harun, R., Arion, F.H., Porutiu, A., Chiciudean, G.O., Todea, A. and Lile, R. (2016) Local residents' attitude toward sustainable rural tourism development. *Sustainability* 8, 100.
Rhiney, K., Walker, T. and Tomlinson, J. (2015) *Strengthening Agritourism Potential in the Caribbean (Study on Agribusiness Development)*. Wageningen: Technical Centre for Agricultural and Rural Cooperation and The Inter-American Institute for Cooperation on Agriculture. See https://www.researchgate.net/publication/282785500_Strengthening_Agritourism_Potential_in_the_Caribbean (accessed 2 November 2019).
Rogerson, C.M. and Rogerson, J.M. (2014) Agritourism and local economic development in South Africa. *Bulletin of Geography. Socio-economic Series* 26, 93–106.

Roszkowska-Mądra, B. and Roman, M. (2015) Znaczenie turystyki wiejskiej w zrównoważonym rozwoju terenów peryferyjnych (na przykładzie gminy Płaska). *Roczniki Naukowe Stowarzyszenia Ekonomistów Rolnictwa i Agrobiznesu* 17 (4), 261–266.

Sammel, A. (2016) Rola agroturystyki w zrównoważonym rozwoju obszarów wiejskich w Polsce. *Współczesne Problemy Rozwoju Turystyki* 6, 77–88.

Suntikul, W., Bauer, T. and Song, H. (2009) Pro-poor tourism development in Viengxay, Laos: Current state and future prospects. *Asia Pacific Journal of Tourism Research* 14 (2), 153–168. doi:10.1080/10941660902847203

Turnock, D. (1999) Sustainable rural tourism in Romanian Carpathians. *The Geographical Journal* 165 (2), 192–199.

UNEP and UNWTO (2005) *Making Tourism More Sustainable: A Guide for Policy Makers*. Nairobi and Madrid: United Nations Environment Programme and World Tourism Organization. See https://wedocs.unep.org/handle/20.500.11822/8741 (accessed 15 May 2019).

WCED (World Commission on Environment and Development) (1987) *Our Common Future*. Oxford: Oxford University Press.

Woś, B., Gawlik, A. and Drobek, W. (2018) Obiekty hotelarskie na terenach wiejskich – nowa jakość usług. *Studia Obszarów Wiejskich, tom 49: Społeczno-kulturowe gospodarcze i przestrzenne, wymiary procesów odnowy i rewitalizacji wsi w Polsce* (pp. 173–185). Warsaw: Komisja Obszarów Wiejskich PTG, Instytut Geografii i Przestrzennego Zagospodarowania PAN.

Xing, Y. and Dangerfield, B. (2011) Modelling the sustainability of mass tourism in island tourist economies. *Journal of the Operational Research Society* 62 (9), 1742–1752. doi:10.1057/jors.2010.77

Zaremba, D. (2008) *Ekoturystyka i odkrywanie dziedzictwa – zbiór dobrych praktyk*. Kraków: Wydawnictwo Fundacja Partnerstwo dla Środowiska.

Zaremba, D. (2009) *Ekoturystyka*. Warsaw: Wydawnictwo Naukowe PWN.

8 Causal Nexus between a Sustainable Economy and Agricultural Credit: Global Evidence

Rashedul Hasan, Mohammad Dulal Miah and Muhammad Ashfaq

This chapter examines the relationship between agricultural credit and economic growth. We have used a vector autoregressive model for testing the Granger causality and identifying the direction of causality between economic growth and agricultural credit after employing a unit root test to verify the stationary consideration for the selected variables. An Augmented Dickey–Fuller (ADF) test provides evidence that variables are stationary at their levels, and thus they are naturally cointegrated. We find evidence of unidirectional causality between agricultural credit and economic growth. This study provides valuable findings for emerging economies and specifies policy recommendation by endorsing the role of central banks to implement an agriculture-friendly credit policy to maintain sustainable economic development.

1 Introduction

The agricultural sector is the most important sector of any economy. It produces consumer products for the survival of human beings. This sector is considered the backbone of all countries, especially developing countries. In the early stages of the development and advancement of societies, the agricultural sector played a fundamental role by increasing production, saving and capital formation, and increased international trade among different countries (Johnston & Mellor, 1961). The importance of agriculture in economic development and uplifting the standards of living of an individual can be discussed through various lenses. For instance, Theodore Schultz, who received the 1979 Nobel Prize for Economics, stressed the importance of agriculture in alleviating poverty. One of the goals of economic development is poverty alleviation through improving the agricultural sector. In economics, poverty has different meanings, and contextual factors need to be understood while making comparisons. One can define poverty as an individual or a household being unable to meet the minimum requirement of food to sustain livelihood in a society (Ogwumike, 2001). One measure of poverty is to assess how much money poor people spend on products and services. The global benchmark was set at US$1.25 per day between 2008 and 2015. However, the poverty line was revised to US$1.90 per day using the most accurate price levels applicable in the revision process (World Bank, 2016). Thus, an individual is considered as being above the poverty line if he or she can spend US$1.90 per day.

To maintain the pace of well-being of an individual, a person who has been alleviated from poverty should be empowered to overcome poverty for a longer time period rather than just for a short while. Such economic development could be achieved by helping them to become an economically sustainable person who can add value to the overall economic development in a society. Mellor (2000) argues that if there is a decrease in poverty, it is because of increasing production in the agricultural sector at the initial stage of development. This increase in agricultural production ultimately leads to structural change in the economy, and to the next stage where industrial and services sectors start to increase their contribution to domestic production. As the productivity of the individual increases, people move from rural to urban areas in search of better opportunities. As a result of this transformation, an economic structural shift from the agricultural sector to the manufacturing sector and eventually to the services sector takes place.

There has always been an increase in demand for agricultural products due to the increasing population. India and China are the major food-consuming regions owing to their large populations. It is expected that the overall agricultural output and fish production will increase by up to 20% in different regions. It is worth reckoning that production growth is expected to be relatively low in the developed world, especially in Western Europe.

On the other hand, Tilman *et al.* (2002) argue that the demand for agricultural products will double in the next 50 years, assuming that an economy wants to maximise the benefits of the agricultural sector. In that case, it must analyse the cost and benefit of alternative agricultural practices by considering the ethical aspects in parallel, because economic policies would become more capital intensive by intensive use of technology, which would lead to structural unemployment. This unemployment will cause a decrease in the demand side of the economy that will have a multiplier effect on other sectors of the economy and, ultimately, it will affect the manufacturing and services sectors.

It is expected that by 2027 South and East Asia, the Americas and Sub-Saharan Africa will witness growth in agriculture and fisheries

(OECD-FAO, 2018). Thus, the agricultural sector should maintain an important role as the population grows. However, trade wars and the resulting rise in agricultural tariffs, especially between China and the United States, could have adverse effects on agricultural production. In the agricultural sector, technology is one of the most important driving factors for innovation and value addition in the output process. Technology plays a significant role in developing the economy to increase labour productivity and is an important factor in increasing the income of the farmer. This increasing marginal productivity raises the demand for technological products that ultimately has a positive spillover effect on the manufacturing and services sectors, as the demand for services provided by the services sector increases in parallel to economic progress.

The supply of agricultural products from a rural area to an urban area requires infrastructure and transport facilities. Transportation and infrastructure play an essential role in the development of the economy, as other industries are involved in building infrastructure. The great benefit-to-cost ratio urges the policymakers to invest more resources in the construction of infrastructure. It ultimately leads to the development of the services sector, as technically skilled staff and financial analysts are major stakeholders of the services sector (Kydd & Dorward, 2001). In net, the service sector has a small proportion of GDP at the initial level but ultimately becomes a significant contributor to GDP. Both the agricultural and industrial sectors increase production in the economy. However, the share of the agricultural and industrial sectors in total GDP decreases eventually compared to the services sector, as developed countries have the most significant proportion of the services sector in total GDP.

The availability of credit for farmers and its impact on the development of the agricultural sector has attracted research attention recently. There are two sources of credit available for farmers, divided into institutional and non-institutional sources. The non-institutional sources consist of money lenders, traders, friends and family, which are often preferred by the farmers due to their inability to secure finance from institutional or formal sources of credit. Our study contributes by examining the causal relationship between agricultural credit disbursed by the formal sector and its impact on economic growth. The rest of the study is divided into four sections. In Section 2 we provide evidence on the importance of agriculture towards economic development by surveying the extant literature. Section 3 discusses the data and methodology, followed by a discussion on empirical findings in Section 4. Finally, we conclude the study with policy implications and by providing directions for future research.

2 Literature Review

Empirical evidence on which the classical theory on agriculture and development often relies can be traced back to the Industrial Revolution.

The Industrial Revolution was preceded by agricultural growth in some countries, including England, Japan and the United States. For instance, the United States employed about three-quarters of its labour force in agriculture, which contributed more than half of the GDP in 1800 (Alvarez-Cuadrado & Poschke, 2011). In 2000, less than 2% of the labour force remained in the agricultural sector, and the share of agricultural production in GDP fell below 2%. Over these two centuries, US output per capita increased more than 25 times. Japan employed about 65% of its labour force in agriculture even as late as 1900 when modern industry had begun to expand substantially. At the same time, agricultural contribution to GDP accounted for 38%. The share of agriculture to GDP declined to a minuscule 1.5%, whereas the share of the labour force employed in the sector plummeted to 5%.

This structural school further draws evidence from the rise of Asian tigers, including South Korea and Taiwan. Lipton and Longhurst (2011) argue that this classical paradigm was successfully implemented in Asia in the 1960s and 1970s, where the Green Revolution averted major famines and simultaneously facilitated industrialisation. Their development was arguably facilitated by land reforms leading to a rapid productivity increase in smallholder farming. The basic premise of the school is that a dynamic transformation from a traditional agriculture-dominant economy to the modern industrial society is feasible if agriculture is modernised through modern technology, which in turn leads to increased productivity. The rise will see the effect of an increase in productivity of the purchasing power of the agrarian population. The demand for non-farm products will rise as a result. Increased demand for non-farm products will pave the way for industrialisation. This hypothesis is labelled as the 'demand-driven' strategy of development (Adelman, 1984). Adelman stresses that the role of increased agricultural productivity via technological innovation and increased investment in raising rural incomes for 'agricultural-demand-led-industrialization' (ADLI) is critical for agro-based economies.

Given the institutional link of the agricultural sector to the rest of the economy, a stimulus to agricultural activities is expected to induce strong demand and, thereby, foster industrial expansion. Vogel (1994) applies the Social Accounting Matrix (SAM) – aggregate structural interrelationships among the various agents of an economy – to identify the importance of agriculture on industrialisation, and finds that, at a low level of development, agriculture possesses strong backward links to non-agricultural activities, possibly so high that a $1 expenditure in agriculture generates $2.75 in induced demand in the non-agricultural sector. The link continues to increase during development. The surplus, linkage and market contributions of agriculture, and the associated multiplier effects, could accelerate growth in the rest of the economy above the rate of growth in agriculture, leading to a relative decline of the agricultural sector in employment and GDP. The role of agriculture in development was thus measured in its support to the acceleration of growth in the rest of the economy, principally industry, with the relative decline of agriculture.

Despite its glorious role in the early stages of industrialisation of advanced nations, the role of agriculture in the classic development paradigm was neglected for about 20 years (De-Janvry, 2010). During the period 1985–2005, agriculture was seen as a sunset industry and not competitive for public investment. This resulted in a sharp decline in public investment and overseas development assistance going to agriculture, and the dismantling of many of the support agencies to agriculture, including agricultural development banks and parastatal marketing agencies. Besides, agriculture was no longer seen as the most effective approach to achieving industrialisation, which could better be obtained through Open Economy Industrialisation strategies based on international financial capital movements, foreign direct investment, international transfers of technology and economies of scale in urbanisation (Rodrik, 2006). De-Janvry (2010) attributes this demise to the drawback of the new development paradigm, which fails to conceptualise the role of agriculture in development adequately.

Moreover, the success of agriculture should be seen beyond what the market can achieve. For example, the conventional measure of development rarely considers such factors as poverty reduction, sectoral disparities, gender equity and intergenerational sustainability in resource use. These parameters are developed more often than not with the development of agriculture, which entails that social choices must be made in weighing up various development outcomes of agriculture, not merely its contribution to wealth accumulation (De-Janvry, 2010).

Up to this point, agriculture has been able to attract widespread attention from scholars and policymakers in terms of the question of how sustainable development of developing economies can be achieved and sustained at a time when rural–urban disparities have widened, the gap between the rich and poor has heightened and environmental degradation has been at a peak. Lo (2012) examines Chinese economic growth through the lens of the Marxian egalitarian pattern of income distribution. The growth acceleration of the Chinese economy so far has been facilitated by two necessary conditions. Firstly, the process of structural change involved both an expansion of the share of industry in the economy and the leading role of a wide range of mass-production industries. Secondly, the egalitarian pattern of income distribution, which underpinned mass consumption, has induced investment and overall expansion of demand (Suzuki & Miah, 2017). Low's analysis throws a clear light on the fact that sustainable development of the Chinese economy critically depends on the equitable distribution of income. Economic acceleration of a growing economy is prone to disruption if the bottom of the population pyramid remains economically destitute.

Chakravarty (1979) scrutinises strategies for Indian economic growth. He favours the agriculture-first strategy as a sustainable approach for industrialisation. He argues that the desirability, as well as the feasibility, of an export-led industrialisation strategy in a slow-growing economy will not be realised unless the agricultural sector is modernised. The long-term growth must primarily be based on the expansion of internal rather than external demand. Similarly, Storm (1995) supports a policy that an increase in agricultural production and income is likely to improve industrial performance because they expand the size of the domestic market. Thus, the industrialisation of countries with a low economic base can mainly be sustained by the expansion of the agricultural sector.

The green revolution in agriculture gives a strong stimulus to non-farm employment through consumption linkage (Hossain, 1988). Modern large-scale agriculture requires industrial products as inputs and raw materials, which means that the demand for service and non-farm products increases, leading to the creation of new job opportunities in the non-farm sector (Alauddin & Tisdell, 1995). At the same time, the wage rate in the non-farm sector increases. An accumulation of capital will follow increased income in both farm and non-farm sectors brought about by the green revolution. Such capital accumulation facilitates capitalist development, leading to the rise of investment in agriculture and industry.

Most underdeveloped economies have failed to make this transition due to various structural obstacles. William Arthur Lewis' two-sector model can help identify these obstacles. Lewis (1954) argues that most developing economies can be characterised by the presence of a dominant agricultural or traditional sector and a small capitalist class. The traditional sector usually absorbs a considerable labour force, which works even at subsistence wage rates. The overall population is so large relative to the capital and natural resources that marginal productivity of labour is insignificant or zero. This induces rural employers to employ more labour at a subsistence wage and expand the sector as long as subsistence labour is available. As a result, the traditional rural sector is incapable of generating reproducible capital as well as capital accumulation. On the other hand, subsistence wages restrict workers' capability to save for non-farm consumption and thereby improve their situation from subsistence to surplus level. This situation is labelled a 'vicious-circle' or low-level equilibrium trap.

In contrast, the urban capitalist sector is tiny and growing. Capitalists recruit labour until the marginal wage equals marginal productivity. At this stage, capitalists enjoy huge surplus as profit which can be expanded for further investment. Given the abundance of the labour force in the traditional sector, expansion of industrial activities seems to be more comfortable as long as investible capital is available. Recruiting additional labour force for the industry, however, does not negatively affect agricultural productivity. This process not only accelerates industrialisation but also increases labour productivity in the traditional sector and continues until the labour surplus disappears. From this view, Lewis argues that the centre for economic development is the rapid accumulation which can be facilitated by converting income distribution to saving.

A critical aspect of the agricultural sector in developing the economy is the share of women in the agricultural sector. A study by the United Nations Development Programme (UNDP) shows that in Nigeria women's participation rate is 60–80% in the agricultural sector and they produce two-thirds of agricultural products; unfortunately, women are voiceless in making agricultural policies and decision making (World Bank, 2016). The contribution of females in the agricultural sector is 40% in Latin America and 60–80% in Africa and Asia. The proportion of women in agriculture in some developing countries is more than 50%, but the productivity level of the female is lower than that of the male (Wilde, 1993). The gap between potential productivity per person and female productivity can be narrowed by incorporating technology into the agricultural sector that requires more technical skills rather than physical attributes.

An increase in technology would raise the contribution of females in total agricultural production, as women could handle it quickly. Bryson (1981) concludes that the failure to recognise the contribution of women in the agricultural sector can be overcome by working with rather than against women. The merit of this approach and the recognition of women in agriculture will encourage women to participate in the sector. Varangis (2015), the Global Lead for Agriculture Finance at the World Bank Group, states that despite the increasing contribution of women in the growth of the agricultural sector in different regions, women face a lack of access to formal financial markets. There are three major barriers to providing adequate finance to women, including lack of market research to understand their needs, poor understanding of the right products and services, and issues in strategy implementation for agricultural finance for women.

3 Data and Methodology

We have used the annual time series data of Gross Domestic Product (GDP) per capita as a proxy to measure economic growth. Data on agricultural credit are taken from the database of the Food and Agricultural Organization of the United Nations. Credit to Agriculture, Forestry and Fishing is used as a proxy for agricultural credit. Both GDP and agricultural credit data are presented in million US$. Our data range between 1991 and 2017, covering 27 years, and include 121 countries. Tables 8.1 and 8.2 provide a summary of agricultural credit and GDP data across regions and income group.

Table 8.1 indicates that the maximum amount of agricultural credit is distributed in the South Asian region, while the minimum agricultural

Table 8.1 Agricultural credit and GDP across the region

Regions	Agricultural credit (million US$)				GDP per capita (million US$)			
	Mean	Min	Max	SD	Mean	Minimum	Maximum	SD
East Asia & Pacific	8,477.840	0.218	61,804.721	14,409.305	9,858.042	109.693	69,591.061	14,724.701
Europe & Central Asia	6,800.257	0.000	77,617.033	16,256.630	12,334.281	228.906	88,214.691	15,559.051
Latin America & Caribbean	800.185	0.056	10,067.585	1,779.653	7,297.657	717.779	63,479.361	7,989.135
Middle East & North Africa	9,126.632	3.242	78,908.000	18,374.489	13,324.637	138.458	88,564.816	17,672.543
North America	6,627.465	124.957	20,500.886	7,379.460	21,359.154	3,542.644	52,270.828	15,584.654
South America	594.558	256.067	2,050.225	510.711	1,581.353	764.140	3,393.956	895.448
South Asia	6,484.932	1.035	165,564.704	25,856.029	1,415.674	130.751	11,151.092	1,863.200
Sub-Saharan Africa	259.334	0.023	3,020.449	436.258	1,890.816	85.642	21,129.233	3,013.482

credit disbursement is found in South America. However, the average agricultural credit is disbursed in the Middle East and North African region, followed by East Asia and the Pacific. Table 8.2 groups the sample countries based on the income group provided by the World Bank. Lower-middle income countries are found to have the maximum disbursement of agricultural credit, while high-income countries have the maximum average agricultural credit. Such information indicates the importance of agriculture on economic development for both high- and low-income countries. Therefore, we proceed to explore the causal link between agricultural finance and economic development.

We employ augmented Dickey–Fuller (ADF) and Phillips–Perron (PP) tests to determine the stationarity properties of our time series. Unit root tests provide a null hypothesis that the series are non-stationary or contain a unit root. We provide the regression equations for the ADF in Equations 1 and 2 that follow an AR(p), and p lagged difference terms for agricultural credit (ACREDIT) and GDP.

$$\Delta ACREDIT_t = \alpha\, ACREDIT_{t-1} + \beta_1\, \Delta ACREDIT_{t-1} + \beta_2\, \Delta ACREDIT_{t-2} + \cdots + \beta_p\, \Delta ACREDIT_{t-p} \quad (1)$$

$$\Delta GDP_t = \alpha\, GDP_{t-1} + \beta_1\, \Delta GDP_{t-1} + \beta_2\, \Delta GDP_{t-2} + \cdots + \beta_p\, \Delta GDP_{t-p} \quad (2)$$

where $\Delta ACREDIT$ and ΔGDP are the first difference of agricultural credit and GDP, respectively; $ACREDIT_{t-1}$ and GDP_{t-1} are lagged values.

After confirming the stationarity status of the time series, we have checked the existence of a long-term relationship in the examined variables. We have used the Johansen and Juselius method to determine the number of cointegration vectors based on the VAR framework's maximum likelihood estimation. Using trace statistics and maximum eigenvalues, these tests conclude the relationship between variables. The lag order is determined using a modified LR statistic, Final Prediction Error (FPE) and Akaike Information Criterion (AIC).

We examine the causality between agricultural finance and economic development using Granger causality tests. First, Granger causality tests are performed using the vector error correction model (VECM) to establish the causal link. In addition to the VECM model, we have also used the Wald Granger non-causality test in the vector autoregressive (VAR) framework in this study following the recommendations of Toda and Yamamoto (1995) and Yamada and Toda (1998). The VECM model involves the regression of the changes in variables on lagged deviations as expressed in the equation:

$$ACREDIT_t = \sum_{i=1}^{p} \Upsilon_1 ACREDIT_{t-i} + \sum_{j=1}^{q} \delta_1 GDP_{t-i} + \mu_t \quad (3)$$

Table 8.2 Agricultural credit and GDP across income group

Income group	Agricultural credit (million US$)				GDP per capita (million US$)			
	Mean	**Min**	**Max**	**SD**	**Mean**	**Minimum**	**Maximum**	**SD**
High income	10,325.33	0.06	78,908.00	18,999.10	21,285.13	649.40	88,564.82	16,781.97
Upper middle income	1,357.20	0.00	31,670.62	3,877.59	4,943.96	172.04	63,479.36	6,094.24
Lower middle income	3,132.74	0.22	165,564.70	14,110.50	3,040.80	109.69	46,356.90	6,589.62
Low income	377.39	0.02	6,472.08	1,020.24	992.66	85.64	21,129.23	2,647.25

$$GDP_t = \sum_{j=1}^{q} \delta_1 GDP_{t-i} + \sum_{i=1}^{p} \Upsilon_1 ACREDIT_{t-i} + \mu_t \tag{4}$$

where $ACREDIT_t$ and GDP_t are as defined earlier; μ_t represents the white noise; p and q are the optimal lags for agricultural finance and economic development, respectively. The VAR model in the bivariate framework is given in Equations 5 and 6.

$$ACREDIT_t = \lambda_0 + \sum_{i=1}^{l+r} \sigma_{1i} ACREDIT_{t-i} + \sum_{i=1}^{l+r} \beta_{1i} GDP_{t-i} + e_{1t} \tag{5}$$

$$GDP_t = \lambda_0 + \sum_{i=1}^{l+r} \sigma_{2i} GDP_{t-i} + \sum_{i=1}^{l+r} \beta_{2i} ACREDIT_{t-i} + e_{2t} \tag{6}$$

where l is the optimal lag order; r is the order of integration; σ and β are the coefficients of *ACREDIT* and *GDP*; and e_1 and e_2 are the white noise error terms. Equations 3 and 5 provide the null hypothesis that *ACREDIT* does not Granger cause GDP. In Equations 4 and 6, the null assumption is that *GDP* does not Granger cause *ACREDIT*.

4 Empirical Results

The results of ADF and PP tests for unit root indicate that both agricultural credit and economic development, proxied by GDP, are stationary at level (Table 8.3). After examining the stationary properties, we selected the optimum lag using Log-Likelihood (LR), Final Prediction Error (FPE) and Akaike Information Criterion (AIC). Table 8.4 reflects that the optimum lag for agricultural credit and GDP is four given all three criteria.

We establish a long-run relationship between the two series using the Johansen cointegration test. The results in Table 8.5 indicate that agricultural credit and GDP per capita have a long-run relationship for the two periods as the eigenvalues are higher than critical values. This indicates a stable long-run relationship between agricultural credit and economic growth. Given the long-run causal relationship between agricultural credit and economic growth, we have conducted a Granger causality test using a VECM and Wald test. Table 8.6 provides results for both Granger causality tests. We are in favour of the null hypothesis that agricultural credit does not Granger cause GDP. However, we reject the null hypothesis that GDP per capita does not Granger cause agricultural credit. Such

Table 8.3 Results of ADF and PP test for unit root at the level

Variable	Test statistics ADF	Test statistics PP	5% Critical value
Agricultural credit	–4.388	–4.450	–2.863 (–2.863)
GDP	–9.679	–9.602	–2.862 (–2.862)

Note: Figure in the parenthesis is a critical value from PP test.

Table 8.4 Lag selection criteria

Lag order	LR	FPE	AIC
0	–	17.718	8.550
1	7514.012	0.033	2.242
2	15.992	0.032	2.278
3	2.573	0.032	2.300
4	19.219*	0.031*	2.307*

Notes: * and ** represent significance at 5% and 1% level, respectively; sample: 1991–2017; number of observations 1193 after adjustments.

Table 8.5 Johansen test for cointegration with restricted constant and no trend in VAR

Null hypothesis	Maximum eigenvalues	5% Critical value	*p*-value	Trace statistic	5% Critical value	*p*-value
$r = 0$	34.526	14.264	<0.001	35.964	15.494	<0.001
$r \leq 0$	1.438	3.841	2.304	1.483	3.841	0.2304

Table 8.6 Granger no causality test

	Granger causality		Block exogeneity Wald test	
Null hypothesis	**Observation**	***F*-statistic**	**Observation**	***Chi*-sq**
Agricultural credit ⇒ GDP	1622	0.969	1858	2.263
GDP ⇒ Agricultural credit	1622	9.549**	1858	25.26**

Note: * and ** represent significance at 5% and 1% level, respectively.

results are consistent with the findings of Sharma (2016) and Alhassan and Fiador (2014).

5 Discussion

The ADF test indicated that variables do not have a unit root at the level. Therefore, all variables included in the empirical analysis are proven stationary. Furthermore, the series of agricultural credit and GDP growth are cointegrated, which proves a long-run relationship. Such results further strengthen the importance of agriculture for the economic growth of a developing country. We provide global evidence of a long-run relationship between agricultural credit and economic development. Anthony (2010) concludes that agricultural credit is an effective instrument for improving agricultural output, which stabilises the GDP. We provide further justification for such findings with a global sample. Table 8.6 indicates that countries with higher mean GDP scores have a larger mean agricultural score, which further validates the Johansen cointegration test results. However, we make an original contribution through the Granger no causality test. Results from the Granger no causality test allow us to uncover that GDP growth has a causal relationship with agricultural credit. Sharma (2016) provide empirical evidence of a unidirectional relationship between financial inclusion and economic growth. While Hartarska *et al.* (2015) find a positive association between agricultural lending and GDP growth, we conclude that such a relationship is bi-directional and GDP growth causes a growth in agricultural credit. Given the institutional link of the agricultural sector to the rest of the economy, such results can justify the reliance of developed economies on the agricultural sector while the sector has become the major contributor for developing economies.

6 Conclusion

The green revolution plays a vital role in the economy. It increases land productivity as the productivity of land increases, and technology brings efficiency in the sowing and harvesting process which increases the farmer's income. This increase in production and decrease in prices has spillover effects on other sectors of the economy, as agricultural products are the raw materials for many industries, such as cotton and jute being the raw materials for the textile industry. On the other hand, as a farmer's income increases, the demand for non-agricultural products also increases. The more income a farmer has, the more money individuals can use and save and, ultimately, the more non-agricultural goods and financial services are demanded. This increasing demand for industrial products persuades industrialists to build new firms to fulfil the demand for non-agricultural products. It allows financial institutions to develop financial products according to the requirements of the rural consumer so that financial transactions can take place smoothly and efficiently. Granger causality results indicate that there is evidence of a unidirectional relationship between agricultural credit and economic development. We do not find agricultural credit to Granger cause GDP growth. Such results can be explained by the decline of public investment and overseas development assistance going to the agricultural sector and drawbacks of the new development paradigm through the emergence of the service sector. Agriculture is no longer seen as the most effective approach to achieve industrialisation. However, we conclude that GDP growth impacts agricultural credit and confirms that the implication of growth causes credit to flow to the agricultural sector. Such results have clear policy implications as adequate and timely inflow of both production and investment credit for the development of the agricultural sector through further reforms would go a long way in the sustained growth of agriculture and food security. Future studies can focus on alternative financing sources for the agricultural sector, with a particular focus on alleviating rural poverty.

References

Adelman, I. (1984) Beyond export-led growth. *World Development* 12 (9), 937–949.

Alauddin, M. and Tisdell, C. (1995) Labor absorption and agricultural development: Bangladesh's experience and predicament. *World Development* 23 (2), 281–297.

Alhassan, A.L. and Fiador, V. (2014) Insurance-growth nexus in Ghana: An autoregressive distributed lag cointegration approach. *Review of Development Finance* 4 (2), 83–96.

Alvarez-Cuadrado, F. and Poschke, M. (2011) Structural change out of agriculture: Labor push versus labor pull. *American Economic Journal: Macroeconomics* 3 (3), 217–258.

Anthony, E. (2010) Agricultural credit and economic growth in Nigeria: An empirical analysis. *Business and Economics Journal* 14 (1), 1–7.

Bryson, J.C. (1981) Women and agriculture in sub-Saharan Africa: Implications for development (an exploratory study). *The Journal of Development Studies* 17 (3), 29–46.

Chakravarty, S. (1979) On the question of home market and prospects for Indian growth. *Economic and Political Weekly* 14 (30/32), 1229–1242.

De-Janvry, A. (2010) Agriculture for development: New paradigm and options for success. *Agricultural Economics* 41, 17–36.

Hartarska, V., Nadolnyak, D. and Shen, X. (2015) Agricultural credit and economic growth in rural areas. *Agricultural Finance Review* 75 (3), 302–312.

Hossain, M. (1988) *Credit for Alleviation of Rural Poverty: The Grameen Bank in Bangladesh*. Washington, DC: International Food Policy Research Institute.

Johnston, B. and Mellor, J. (1961) The role of agriculture in economic development. *American Economic Review* 87 (2), 566–593.

Kydd, J. and Dorward, A. (2001) The Washington consensus on poor country agriculture: Analysis, prescription and institutional gaps. *Development Policy Review* 19 (4), 467–478.

Lewis, A. (1954) Economic development with unlimited supplies of labour. *Manchester School of Economic and Social Studies* 22 (2), 139–191.

Lipton, M. and Longhurst, R. (2011) *New Seeds and Poor People*. London: Routledge.

Lo, D. (2012) *Alternatives to Neoliberal Globalization: Studies in the Political Economy of Institutions and Late Development*. Basingstoke: Palgrave Macmillan.

Mellor, J.W. (2000) Agricultural growth, rural employment, and poverty reduction: Non-tradables, public expenditure, and balanced growth (Agriculture and Rural Development Discussion Paper). Washington, DC: World Bank. See http://documents.worldbank.org/curated/en/649461468766828655/pdf/multi0page.pdf (accessed 11 October 2019).

OECD/FAO (2018) OECD-FAO Agricultural Outlook 2018–2027, OECD Publishing, Paris/Food and Agriculture Organization of the United Nations, Rome. See https://doi.org/10.1787/agr_outlook-2018-en

Ogwumike, F.O. (2001) Profile and dimension of poverty in Nigeria. In *NCEMA Workshop on Poverty Reduction*. Ibadan: Development Policy Centre.

Rodrik, D. (2006) *Industrial Development: Stylized Facts and Policies Revised*. Cambridge, MA: Harvard University Press. See http://j.mp/2vw59Ud (accessed 10 December 2019).

Sharma, D. (2016) Nexus between financial inclusion and economic growth: Evidence from the emerging Indian economy. *Journal of Financial Economic Policy* 8 (1), 13–36.

Storm, S. (1995) On the role of agriculture in India's longer-term development strategy. *Cambridge Journal of Economics* 19 (6), 761–788.

Suzuki, Y. and Miah, M.D. (2017) China's 'new normal': Explaining China's economic development from institutional and Marxian perspectives. In A. Banik, M.K. Barai and Y. Suzuki (eds) *Towards a Common Future: Understanding Growth, Sustainability in the Asia-Pacific Region* (pp. 21–41). Singapore: Palgrave Macmillan.

Tilman, D., Cassman, K.G., Matson, P.A., Naylor, R. and Polasky, S. (2002) Agricultural sustainability and intensive production practices. *Nature* 418 (6898), 671.

Toda, H.Y. and Yamamoto, T. (1995) Statistical inference in vector autoregressions with possibly integrated processes. *Journal of Econometrics* 66 (1/2), 225–250.

Varangis, P. (2015) *Financing to Support Women in the Agricultural Sector.* Washington: World Bank Group.

Vogel, S.J. (1994) Structural change in agriculture: Production linkages and agricultural demand-led industrialization. *Oxford Economic Papers* 46 (1), 136–156.

Wilde, V.L. (1993) Agricultural extension and farm women in the 1980s. Food and Agriculture Organization of the United Nations, Rome.

World Bank (2016) *Principles and Practice in Measuring Global Poverty.* See https://www.worldbank.org/en/news/feature/2016/01/13/principles-and-practice-in-measuring-global-poverty (accessed 7 November 2019).

Yamada, T. and Toda, H.Y. (1998) Inference in possibly integrated vector autoregressive models: Some finite sample evidence. *Journal of Econometrics* 86 (1), 55–95.

Part 2

Poverty and Development: A Local Perspective

9 Measuring Sustainable Development: A Quantitative Evaluation of Sustainable Development in Vietnam

Van Canh Truong

Sustainable development requires a balance between three main components – social, economic and environmental – to meet the needs of the present without compromising the ability of future generations to meet their own needs. Development of the evaluation of relevant indicators and the method for assessing the level and status of sustainable development enables effective monitoring and measurement of progress towards sustainable development. The research aim was to construct a comprehensive index by using indicators to measure and evaluate the changes in the level of sustainable development in Vietnam between 2004 and 2016. Based on the definition of sustainability, the criteria for selecting indicators and the available data for Vietnam, a system of 23 core indicators that address the three pillars of sustainable development was created. This chapter outlines the calculation methodology used to develop a comprehensive index for different times during the 2004–2016 period. The research indicates that there were significant improvements in the value of a synthetic indicator for sustainable development of Vietnam, from 0.497 in 2004 to 0.588 in 2016. However, these scores also indicate a low–medium level of sustainable development in Vietnam and, in particular, a low level for the economic and environmental components.

1 Introduction

The present generations live better than those that came before us because we have inherited the benefits of their decisions, discoveries and achievements. However, we now face many challenges that threaten future outcomes, such as climate change, poverty and pollution. To ensure that our children and grandchildren can maintain a good quality of life and have a higher standard of living, we must leave future generations a good social, economic and environmental foundation (Australia National Sustainability Council, 2013: 267). To do this, our progress must be made in sustainable ways.

Sustainable development is a holistic approach linking economic development, social inclusion and environmental sustainability (Sachs, 2015). It attempts to balance economic growth, environmental protection and social progress. Nowadays, sustainable development is a strategic choice selected by all nations. Establishing an efficient, healthy and civilised society, including liveable human settlements and regional ecological security, has become a priority for governments and institutions around the world. To reach the targets of sustainable development, the international community has launched many summits and conferences to discuss the issue and has also released a number of indicator and index systems to monitor, measure and evaluate the progress towards sustainability. However, these systems cannot be fully applied to the unique circumstances of some countries. Therefore, development of a quantitative method to measure the level of sustainable development, followed by its application in a case study, is necessary.

Vietnam is experiencing a period of rapid urbanisation and industrialisation, accompanied by an increasing Gross Domestic Product (GDP), expanding infrastructure and a change in the quality of life. Even though there are many severe problems and shortcomings that must still be dealt with, Vietnam has recorded remarkable achievements in economic, social and environmental sectors, the three pillars of sustainable development (Vietnam Socialist Republic, 2008: 84). Generally speaking, the significance of sustainable development has already been recognised and some research has been carried out. But there are no useful evaluation methods nor a comprehensive index for the assessment of sustainable development in Vietnam.

The next section of this chapter defines the term 'sustainable development'. The method of measuring sustainability, along with a new set of core indicators developed to be suitable for Vietnam, is then explained. Finally, the method of calculating the indicators, aggregating the results into categories and then into one index for the entire country from 2004 to 2016, is illustrated.

2 The Definition of Sustainable Development

The concept of sustainable development appeared for the first time in 1987 with the publication of the Brundtland Report, which warned of the negative environmental consequences of economic growth and globalisation and tried to find possible solutions to the problems caused by industrialisation and population growth (ACCIONA, n.d.). The brief definition of sustainable development in the report is the 'ability to make development sustainable to ensure that it meets the needs of the present without compromising the ability of future generations to meet their own needs' (WCED, 1987). This is undoubtedly the typical definition when judged by its ubiquitous use and frequency of citation (Kates *et al.*, 2005). However, the term 'sustainable development' has been given many definitions and interpreted in more than 200 ways (Rogers *et al.*, 2008; van de Kerk & Manuel, 2008). An enormous number of books, articles, reports and policy papers offer many proper definitions of the term (Gibson *et al.*, 2005).

For many people, the basic idea of sustainable development focuses mainly on the depletion of resources. Others consider that sustainability also covers pollution, conservation of nature and other environmental and ecological aspects (van de Kerk & Manuel, 2008). Some people see sustainable development as primarily about the environment, and indeed sustainable development has its roots in conservation and the desire to protect the planet's ecosystems from the ravages of human civilisation (Blewitt, 2018).

However, sustainable development does not mean the sustainable development of any single economic, social or environmental subsystem alone. Today, sustainable development is broadly understood as a three-pillar approach that attempts to balance and seeks to reconcile three complex systems: the economy, the society and the environment (Baker, 2006; Flint, 2013; Li *et al.*, 2009; Rogers *et al.*, 2008). 'Balance', 'equilibrium', 'harmony' and 'reconcile' are all words used in the literature to describe the idea of the balance in development between different systems on earth. Sustainable development seeks to build a good society – not only an economically prosperous society (with a high per capita income) but also one that is socially inclusive and environmentally sustainable. It requires balanced and holistic development in which economic progress is widespread, extreme poverty is eliminated, social inclusion and cohesion are achieved and the environment is protected from degradation (Sachs, 2015). In this sense, a higher level of sustainability brings us closer to equilibrium, as a high level of economic development goes together with a high level of social well-being and environmental protection.

Moreover, sustainable development requires an understanding of the relationship, the integration and the interactions within and between elements of a system. Sustainable development requires a holistic way of looking at the world and human life (Blewitt, 2018). Sustainable development is not a simple way to achieve a better society. It can be applied to existing plans and programmes, used to assess individual projects or to form the basis for large-scale planning discussions (Cooper & Vargas, 2004). However, we have to perceive that it is a complex system, meaning that sustainable development is not like the sum of its parts, but the interactions of the components, which require a holistic approach and certain complexity of thinking as well (Sachs, 2015).

From all of these perspectives, sustainable development is the development that guarantees the balance between economic development, social well-being and environmental components to satisfy the needs of the present without compromising the ability of future generations to meet their own needs.

This definition shows us that sustainable development is also a way of looking at the world. Firstly, sustainable development focuses on our shared aspirations for a decent life in a balanced, holistic framework, combining economic development, social inclusion and environmental sustainability. Secondly, it is a way of describing the interlinkages of economic, social and environmental change. Therefore, to measure sustainability we have to measure two aspects (the vertical and horizontal directions). In the vertical direction, we need to measure and assess the level of sustainability in all of its aspects in the sense of improvement towards goals of sustainable development in order to access the balance and holistic nature of the development between the economic, social and environmental components. One way to do this is to construct a comprehensive index, which will be revealed in this chapter (see Section 4), and apply it to measure the level of sustainable development in Vietnam. Secondly, in the horizontal direction, we need to measure the level of sustainability by determining the intra-system and inter-system equilibrium or the interrelationships between and within components of sustainable development.

3 Development of an Indicator System for Vietnam

There are now many famous comprehensive indexes and sets of indicators concerning sustainability on a national level designed for use anywhere in the world, such as: the Human Development Index developed by the UNDP; the Environmental Performance Index developed by Columbia University and Yale University, USA; the Ecological Footprint published every two years by the WWF; the Millennium Development Indicators set up by the UN in 2000; and the Sustainable Development Goals set up by the UN in 2015. In Vietnam, the Prime Minister of the Government of Vietnam has signed a decision to release a strategy for the sustainable development of Vietnam in the period 2011–2020, with indicators to monitor and evaluate sustainable development for the whole country.

The research included an examination of the pros and cons of the indexes and relevant existing indicators, keeping in mind the definition and components of sustainability given above, as well as consideration of circumstances in Vietnam. The general conclusion is that none of the existing comprehensive indexes and the relevant set of indicators seems to fit our purposes entirely. In other words, no one method can be used completely to measure sustainability in Vietnam in an easily understandable and suitable way. Accordingly, a new comprehensive index had to be developed, based on a set of indicators following the definition of sustainable development, taking into account the circumstances of Vietnam.

One of the essential steps in the construction of a comprehensive index is to consider the criteria for choosing indicators. An indicator has to be selected carefully, meeting SMART (Specific, Measurable, Available, Relevant, Time-related) criteria as follows (Bell & Morse, 2003; Lynch *et al.*, 2011: 62; Sachs *et al.*, 2016; van de Kerk & Manuel, 2008):

- Specific – indicators have to be independent of each other and must have no overlap. Essentially, we should avoid using indicators that have a high correlation in the same component of sustainable development.
- Measurable – an indicator must be measurable (implies that it must be a quantitative indicator and we can set an upper and lower threshold for it).
- Available – data for the indicators must be available, reliable, from public sources, scientific or official institutional sources and also available for all administrative units of the research.
- Relevant – an indicator must be relevant to an issue according to the definition used.
- Time-related – data must be recent and regularly updated.

A new set of 23 indicators, which takes into account the interpretation of the definition of sustainable development with its three main components, the criteria for the indicators and Vietnam's circumstances, was defined. These indicators, each with data for 2004, 2008, 2012 and 2016, are clustered into three components (see Table 9.1).

Table 9.1 New set of indicators for sustainable development in Vietnam

Component		Indicator	2004	2008	2012	2016	Source
Economic	1	GDP per capita (PPP current US$)	2757.24	3852.18	4909.95	6295.59	World Bank
	2	Urban population rate (%)	26.53	28.99	31.83	34.51	GSO
	3	Proportion of GDP generated by services	37.98	42.51	37.27	40.92	GSO
	4	Incremental capital–output ratio (ICOR)	5.00	6.75	6.76	6.41	GSO
	5	Unemployment rate (% labour force)	2.10	2.38	1.96	2.30	GSO

(continued)

Table 9.1 (*Continued*)

Component		Indicator	2004	2008	2012	2016	Source
	6	Percentage of highly trained employed workers at 15 years of age and above (%)	12.00	14.30	16.60	20.60	GSO
	7	Global Competitiveness Index (1–7)	3.47	4.10	4.11	4.31	World Economic Forum
Social	8	Adult literacy rate (%)	90.30	93.60	94.70	95.00	GSO
	9	Proportion of households that own a permanent house (%)	20.80	27.80	49.60	49.70	GSO
	10	Poverty rate (%) – national poverty line	18.10	13.40	11.10	5.80	GSO
	11	Gini index	0.42	0.43	0.42	0.44	GSO
	12	Mortality rate – under-fives (‰)	29.40	22.50	23.20	21.80	GSO
	13	Proportion of children under five with malnutrition (underweight) – %	26.60	19.90	16.20	13.90	Ministry of Health
	14	Average life expectancy at birth	73.96	74.64	75.32	76.25	World Bank
	15	Proportion of deaths due to traffic accidents per 100,000 population	14.74	13.52	11.08	9.37	GSO
Environmental	16	Forest cover (%)	37.60	38.70	40.70	40.80	GSO
	17	Natural forest per total forest area (%)	82.40	78.90	75.20	72.40	GSO
	18	Agricultural land per person (ha)	0.12	0.11	0.11	0.13	GSO
	19	Proportion of households with access to improved sanitation (%)	61.00	65.00	77.40	83.30	GSO
	20	Percentage of households with access to potable water (%)	80.70	92.10	91.00	93.40	GSO

Table 9.1 (*Continued*)

Component		Indicator	2004	2008	2012	2016	Source
Environmental	21	Proportion of rural households using wood for cooking (%)	70.32	66.48	63.60	33.36	GSO
	22	Annual mean concentration of particulate matter 2.5 (µg/m^3)	21.89	25.65	26.40	27.30	World Bank
	23	Renewable electricity output (% of total electricity output)	38.56	35.48	45.02	41.65	World Bank

Source: Author's calculations and synthesis (GSO and World Bank: Statistical Yearbook of Vietnam 2004–2016; Result of the Vietnam Household Living Standards Survey 2004, 2006, 2008, 2010, 2012, 2014, 2016; Vietnam Population and Housing Census 2009).

4 Method for Constructing a Comprehensive Index

The research started by rescaling each variable from 0 to 1 with 0 representing the worst performance and 1 describing the optimum. This procedure was necessary to make the raw data comparable across indicators. The author was always conscious that for some variables a high numerical value is negative (e.g. infant mortality rate), while for others a high numerical value is positive (e.g. life expectancy at birth). To have rescaled variables we first need to define thresholds, the upper and lower bounds, for each indicator. The thresholds for each indicator were set by looking for the best and worst performances of almost all countries in the world in order to be able to compare the achievements of Vietnam on the road to sustainable development with other countries and the world.

In most cases, the absolute thresholds could be defined as optimum natural values to mark the upper bound for the variable. These were derived from maximum values that must be met to achieve sustainable development. For example, in the case of poverty or undernourishment, 0% is optimal, and in some instances, such as the adult literacy rate or households using electricity, 100% is optimal.

For some variables, no absolute upper bounds could be identified in this way as it may be impossible to achieve certain absolute limits. For instance, it is impossible and utopian to achieve a zero Gini index, zero child mortality rate or zero deaths from road accidents. In these cases, the research took the average of the five best performers in the world as the upper threshold (Sachs *et al.*, 2016). It is essential to bear in mind that the threshold value cannot always be determined objectively, nor will it ever be a constant value over time (van de Kerk & Manuel, 2008).

To determine the lower threshold, data were truncated by removing the bottom two and a half percentiles from the distribution in order to remove outliers and the effect of extreme values at the bottom end of the distribution, which can change the results of a composite index. The same weight was used for each indicator, meaning that within each component each indicator was equally weighted. The approach treats all indicators equally and offers the advantage that as data for more and better sustainability indicators become available, new variables can easily be added to the comprehensive index without changing the relative weighting of the components (Sachs *et al.*, 2016).

After establishing the upper and lower bounds, variables were transformed linearly to a scale between 0 (minimum value) and 1 (maximum value). In other words, each indicator, component and synthetic index is always somewhere between the worst state (0) and an optimum state (1) (Liu *et al.*, 2018). The equation below was applied to calculate a value for all of the indicators:

$$I_n = \frac{x_a - x_{min}}{x_{max} - x_{min}}.$$

where I_n is the adjusted value of each indicator, x_a is the raw value, x_{min} the worst performance and x_{max} the best performance.

A logarithm base 10 was applied to this equation when the author counted the value for GDP per capita. Any resulting values above 1 were set to 1, and negative values were set to 0. This formula ensures that higher values denote better performance and all rescaled variables are expressed as ascending variables. The total value for each component was achieved by calculating

$$S_i = \frac{\sum_1^n I_n}{n}$$

where S_i is the value of each component which is formed by indicators (I_n), and n is the number of indicators in each pillar. The comprehensive index was constructed by taking the average of the component values:

$$CI = \frac{S_1 + S_2 + S_3}{3}$$

where CI is the Comprehensive Index, S_1 is the economic sustainable development index, S_2 the social sustainable development index, and S_3 an environmental sustainable development index.

The adjusted value of each indicator, each component and the comprehensive index were evaluated across six levels to determine the strength of capacity for sustainable development (see Table 9.2).

Table 9.2 Classification for sustainable development levels of Vietnam

Level	Value	Qualitative level	Sustainability level
I	0.8–1.0	Very high	Sustainable
II	0.7–0.8	High	Almost sustainable
III	0.6–0.7	High–medium	Medium
IV	0.5–0.6	Low–medium	
V	0.4–0.5	Low	Almost unsustainable
VI	0.0–0.4	Very low	Unsustainable

Source: Author's analysis.

5 A Comprehensive Evaluation of Sustainable Development in Vietnam

The author first calculated the adjusted values for each indicator within the three components (economic, social and environmental) by using the method for constructing the index and the data for indicators for Vietnam from 2004 to 2016 described in Table 9.1. The adjusted results for each indicator of each component are presented in spider charts. The index for each component and a comprehensive index of sustainability for the whole country in the four years 2004, 2008, 2012 and 2016 are presented in a bar chart.

5.1 Economic component

Vietnam has made the transition from a centrally planned economy to a market-oriented system with unprecedented success, and in 2010 the country achieved low middle-income status (Asian Development Bank, 2013: 68). But when we measure the distance to sustainability for the economic component, we see Vietnam remain on the lower rungs of the ladder. In the 2004 data, the synthetic indicator for the economic component in Vietnam was low (0.423), and ranked as Level V (low). It is worth noting that improvement of the level of this component took place slowly, reaching 0.445 in 2008 and 0.456 in 2012, with Vietnam still stuck at Level V (low) even in 2016, with 0.494 (see Figure 9.4, p. 139). The major reasons for this problem were the low per capita GDP, the proportion of GDP generated by services, the urban population rate, the level of competitiveness of the economy and the particularly low percentage of highly trained employed workers. Moreover, ICOR, an indicator used to determine a country's level of production efficiency, was still high and set to increase, meaning that the score for ICOR was going to decrease. The bright spot in the economic component was the sustained low level of the unemployment rate (Figure 9.1). However, it is worth noting the relatively high proportion of unemployment in trained labour (professional training

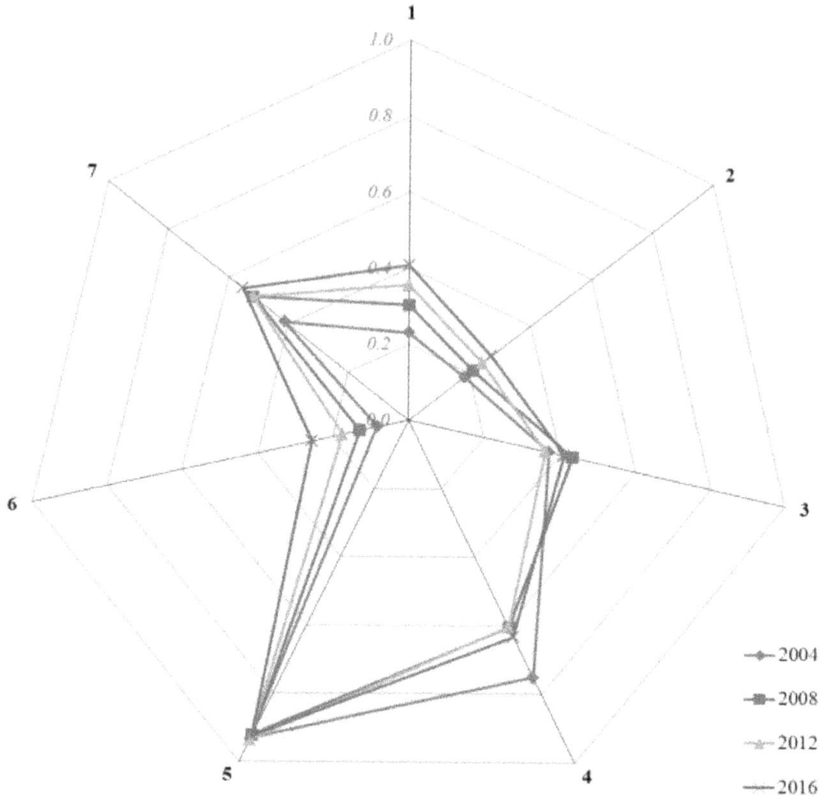

Figure 9.1 Comprehensive evaluation of the economic component in Vietnam
Notes: (1) GDP per capita (PPP current US$); (2) proportion of the population living in urban areas; (3) proportion of GDP generated by services; (4) incremental capital–output ratio (ICOR); (5) unemployment rate (% labour force); (6) percentage of highly trained employed workers at 15 years of age and above; (7) Global Competitiveness index.

meaning a college degree or higher) compared to the total number of unemployed workers (27.2%) (GSO, 2017).

5.2 Social component

The Vietnamese government has implemented many policies that support human well-being. Over the years, Vietnam has made outstanding achievements in education, living conditions, healthcare and poverty reduction (Agenda 21 Office, 2008). These are the main reasons why the social component of the sustainable development of Vietnam was at a good level and had a higher score than other areas during the period studied. Figure 9.4 (p. 139) shows that in 2004 the comprehensive indicator for the social component was 0.574; it then increased significantly to 0.631 in 2008 and 0.693 in 2012 at Level III (high–moderate) and in 2016 it reached

0.720 at Level II (high). The highest scores were in the adult literacy rate, poverty rate, mortality rate for under-fives and the average life expectancy at birth. The Gini index, which is used to assess social equality through income distribution, did not improve significantly. In fact, the gap between the rich and poor is increasing. According to the General Statistics Office of Vietnam, the distance between the highest income quintile and the lowest was 9.2 times in 2010 and increased to 9.8 times in 2016 (GSO, n.d.). The proportion of households that own permanent houses and the proportion of malnutrition in children under five were improving, even though the scores for these indicators were still low. The proportion of deaths due to traffic accidents per 100,000 population was going down significantly, but it should be mentioned that this was still a large number, with an average of 8670 deaths per year due to traffic accidents (GSO, n.d.) (Figure 9.2).

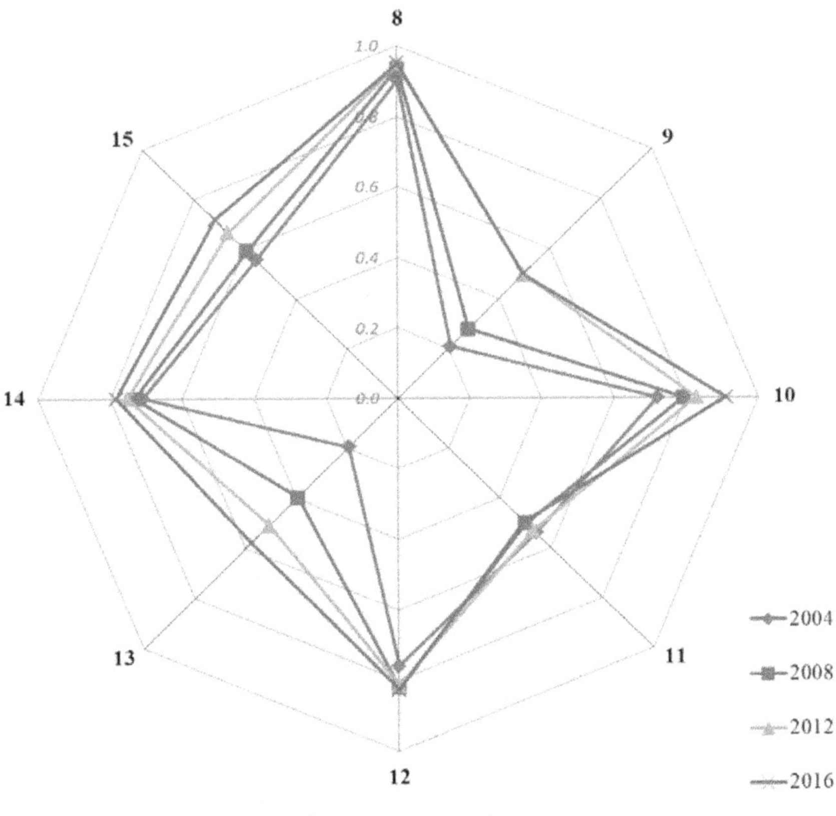

Figure 9.2 Comprehensive evaluation of the social component in Vietnam

Notes: (8) Adult literacy rate; (9) proportion of households that own a permanent house; (10) poverty rate (%) – national poverty line; (11) Gini index; (12) mortality rate – under-fives (‰); (13) proportion of children under five with malnutrition (underweight); (14) average life expectancy at birth; (15) proportion of deaths due to traffic accidents per 100,000 population.

5.3 Environmental component

Analysis showed a low level for the environmental component in Vietnam. The synthetic indicator was poor at 0.494 in 2004 and 0.491 in 2008 (Level V). The index then increased gradually to 0.515 in 2012 and 0.551 in 2016 at Level IV (low–medium) (Figure 9.4, p. 139). Vietnam has made significant improvements in the proportion of households with access to improved sanitation and the percentage of households with access to potable water. The proportion of rural households using wood for cooking went down dramatically to 33.36% in 2016. Yet there remain many challenges for Vietnam in improving the quality of the environment. Natural forest accounted for a high proportion of the total forest area but this decreased gradually. The percentage of forest cover was still low for a country with more than 75% mountainous regions. With just 0.13 ha (2016) agricultural land area per person, Vietnam is still an agricultural country with the vast majority (65%) of the population living in rural areas (GSO, 2017). This reflects the fact that the ecological footprint, which measures human demands on nature, is so big and there is huge pressure exerted on the environment. Improving the air quality has become an urgent concern. The annual mean concentration of particulate matter 2.5 (μg/m^3) in the studied period was high and consistently above the national standard (25 μg/m^3), especially in metropolises such as Hanoi and Ho Chi Minh City (Figure 9.3).

5.4 A comprehensive evaluation of sustainable development

Generally, over the course of the study, Vietnam was still at a low–medium level and moving slowly towards a higher level of sustainability. The comprehensive index for sustainable development for Vietnam shows that the level increased gradually from 0.497 in 2004, a poor level, to Level IV (low–medium) with 0.523 in 2008, 0.555 in 2012 and 0.588 in 2016 (Figure 9.4). It also shows a significant difference in the level of sustainable development across the components. Vietnam achieved the highest level for the social dimension, followed by the environmental and economic components.

6 Discussion and Conclusion

One of the most basic and essential questions is how to measure and evaluate sustainability. I acknowledge that measuring sustainable development is challenging since it covers a wide range of issues. Based on the definition of sustainable development, I devised that to measure sustainability we need to determine the level of sustainability for the whole system and in specific subsystems in a vertical direction, and the inter- and intra-system equilibrium in a horizontal direction, which indicates, in fact, the system's sustainability. In this chapter, I have shown only the method of constructing a comprehensive index to measure the level of

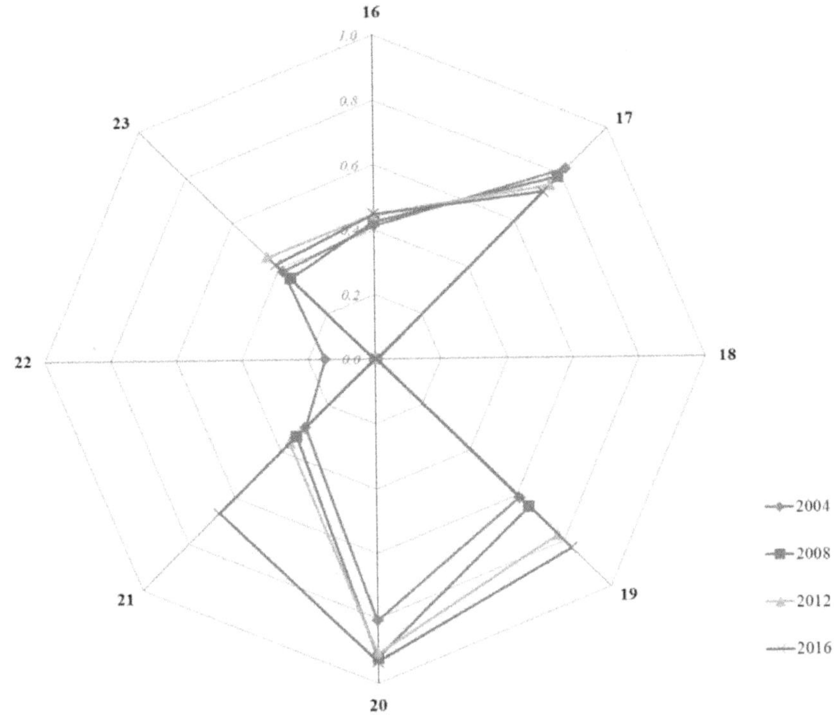

Figure 9.3 Comprehensive evaluation of the environmental component in Vietnam

Notes: (16) Forest cover (%); (17) natural forest per total forest area (%); (18) agricultural land per person (ha); (19) proportion of households with access to improved sanitation; (20) percentage of households with access to potable water; (21) proportion of rural households using wood for cooking; (22) annual mean concentration of particulate matter 2.5 (μg/m³); (23) renewable electricity output (% of total electricity output).

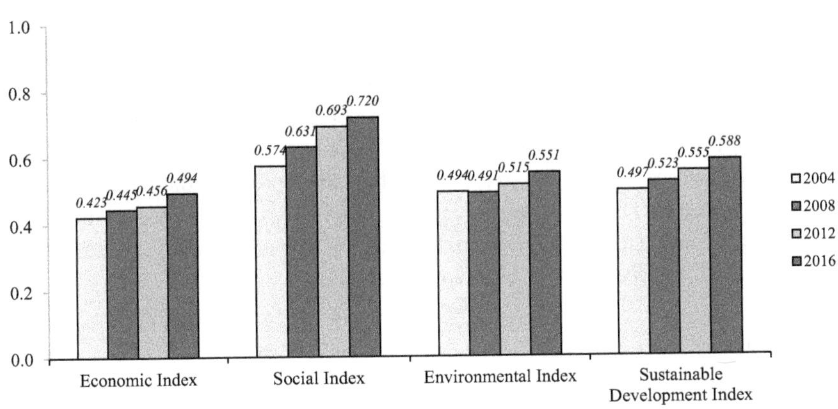

Figure 9.4 Comprehensive evaluation of the sustainable development level of Vietnam

sustainable development in a vertical direction. The method described in this chapter provides a suite of individual indicators that can be combined into a series of comprehensive hierarchical indicators. The study used individual core indicators that can be grouped into three component indicators (economic, social and environmental components), which can in turn be combined into a single first-level comprehensive indicator. An advantage of the approach is that progress towards a goal of sustainable development can be quantified by measuring the indicators. The method combines indicators by using a multidimensional approach that better reflects the integrative system principle that the whole is more than the sum of its parts. The method only requires analysts to set upper and lower limits (by looking for the best and worst performances of almost all countries in the world) rather than employing subjective expert judgements on the values of coefficients, and therefore reduces the level of bias that can distort subjective assessments (Li *et al.*, 2009). Earlier in the chapter I discussed the indicator system that was developed to assess the sustainable development of Vietnam. This is a system of 23 core indicators that address three aspects of sustainable development. The research used the historical and the most recent available data to study the changes across the entire country over time. The main results, the levels of sustainable development for the country from 2004 to 2016, indicate that Vietnam was still at the low–medium level of sustainable development but getting better over some time; there were significant differences in the level of sustainability between the components, with the social pillar higher than others. However, the research still lacks indicators for some essential aspects, especially in the environmental component, such as land quality, environmental conservation and biodiversity preservation, due to a lack of available historical statistical data. Updating relevant indicators with available data is necessary to improve the quality of the research. Looking ahead to the expansion of the research in the future, in addition to measurement by using the comprehensive index, the level of sustainability can be measured by determining the intra-system and inter-system equilibrium of sustainable development or measuring the interconnectedness and relationships within and between components of sustainability. In this sense, the higher the level of interconnectedness within and between subsystems of sustainable systems, the higher the level of sustainability. For this purpose, methods of multivariate statistical analysis can be used, including correlations analysis, principal component analysis, canonical analysis and multivariate regression analysis.

References

ACCIONA (n.d.) *What is Sustainable Development, and What are the Global Goals?* See https://www.acciona.com/sustainable-development/ (accessed 30 January 2019).

Agenda 21 Office (2008) *Sustainable Development Implementation in Vietnam.* Hanoi: Agenda 21 Office – Ministry of Planning and Investment of Vietnam.

Asian Development Bank (2013) *Vietnam: Environment and Climate Change Assessment*. Manila: Asian Development Bank.

Australia National Sustainability Council (2013) *Sustainable Australia Report 2013: Conversations with the Future*. Canberra: DSEWPaC.

Baker, S. (2006) *Sustainable Development*. New York: Routledge.

Bell, S. and Morse, S. (2003) *Measuring Sustainability: Learning by Doing*. London: Routledge.

Blewitt, J. (2018) *Understanding Sustainable Development*. New York: Routledge.

Cooper, P.J. and Vargas, C.M. (2004) *Implementing Sustainable Development: From Global Policy to Local Action*. Lanham, MD: Rowman & Littlefield.

Flint, R.W. (2013) *Practice of Sustainable Community Development: A Participatory Framework for Change*. New York: Springer.

Gibson, R.B., Hassan, S., Holtz, S., Tansey, J. and Whitelaw, G. (2005) *Sustainability Assessment: Criteria, Processes and Applications*. New York: Earthscan.

GSO (General Statistics Office) (2004–2016) *Result of the Vietnam Household Living Standards Surveys*. Hanoi: Statistical Publishing House.

GSO (General Statistics Office) (2006–2016) *Vietnam Rural, Agriculture and Fishery Census*. Hanoi: Statistical Publishing House.

GSO (General Statistics Office) (2010) *Vietnam Population and Housing Census 2009*. Hanoi: Statistical Publishing House.

GSO (General Statistics Office) (2017) *Vietnam Statistical Yearbook 2016*. Hanoi: Statistical Publishing House.

GSO (General Statistics Office) (n.d.) *Statistical Data*. See https://www.gso.gov.vn/en/homepage/ (accessed 27 January 2019).

Kates, R.W., Parris, T.M. and Leiserowitz, A.A. (2005) What is sustainable development? Goals, indicators, values, and practice. *Environment: Science and Policy for Sustainable Development* 47 (3), 8–21.

Li, F., Liu, X., Hu, D., Wanga, R., Yanga, W., Li, D. and Zhao, D. (2009) Measurement indicators and an evaluation approach for assessing urban sustainable development: A case study for China's Jining City. *Landscape and Urban Planning* 90 (3–4), 134–142.

Liu, B., Jiang, X., Jin, T., Li, B., Ma, H., Zheng, Z. and Li, J. (2018) The research on the LDC's sustainable development. *IOP Conference Series: Earth and Environmental Science* 151 (1): 012016.

Lynch, A., Andreason, S., Eisenman, T., Robinson, J., Steif, K. and Birch, E.L. (2011) *Sustainable Urban Development Indicators for the United States* (p. 62). Philadelphia, PA: PENN Institute for Urban Research.

Rogers, P.P., Jalal, K.F. and Boyd, J.A. (2008) *An Introduction to Sustainable Development*. New York: Earthscan.

Sachs, J. (2015) *The Age of Sustainable Development*. New York: Columbia University Press.

Sachs, J., Schmidt-Traub, G., Kroll, C., Durand-Delacre, D. and Teksoz, K. (2016) *SDG Index and Dashboards – Global Report*. New York: Bertelsmann Stiftung and Sustainable Development Solutions Network (SDSN).

van de Kerk, G. and Manuel, A.R. (2008) A comprehensive index for a sustainable society: The SSI – the Sustainable Society Index. *Ecological Economic* 66 (2–3), 228–242.

Vietnam National Institute of Nutrition (n.d.) *National Statistical Data for Health and Nutrition*. See http://viendinhduong.vn/vi/so-lieu-thong-ke.html (accessed 2 February 2019).

Vietnam Socialist Republic (2008) *Sustainable Development Implementation in Vietnam*. Hanoi: Vietnam Agenda 21 Office.

WCED (World Commission on Environment and Development) (1987) *Our Common Future*. Oxford: Oxford University Press.

World Bank (2017) *World Development Indicators*. See https://datacatalog.worldbank.org/dataset/world-development-indicators (accessed 2 February 2019).

10 Evaluation of Consumers' Knowledge, Attitudes and Practices in Sustainable Consumption: A Romanian Case Study

Cristina Ghinea and Ancuța Elena Prisacaru

One of the most important challenges for sustainable development is unsustainable consumption. Nowadays, we are facing climate change, resource depletion, food loss and environmental pollution, to which unsustainable consumption also contributes. In this work, we evaluate the knowledge, attitudes and practices in sustainable consumption of Romanians from the northeast of the country. An online questionnaire was used to collect data. Two hundred and twenty-six people responded to the survey, of whom 179 were females and 47 were males. Results show that only 1.7% of the interviewed females had poor knowledge regarding sustainable consumption, while 88.8% had a high level of knowledge. Females with poor knowledge were aged between 18 and 34, with high school level education, income below the average and the majority from rural areas; 45.2% of the female responders had a moderate attitude, while 64.8% had moderate practices regarding sustainable consumption. Of the males, 91.4% had a high level of knowledge regarding sustainable consumption, and the majority had moderate attitudes and practices (38.2% and 72.3%, respectively). The results also showed that 68.1% of the females and 56.5% of the males considered the rapid development of society to be the main cause of environmental degradation, and 23.9% of male respondents believed that environmental degradation is due to poverty, while only 10% of females gave the same response. However, 21.7% of females and 19.5% of males considered population growth to be the major cause of environmental degradation. The results were also analysed with a one-way analysis of variance (ANOVA) using Minitab software.

1 Introduction

Sustainable development has many definitions, but the most commonly used definition is that presented in the Brundtland Report (1987): 'development that meets the needs of the present without compromising the ability of future generations to meet their own needs'. Social, economic and environmental priorities are part of this complex process, which is meant to improve people's well-being (Tkalec, 2010; Wesarat *et al.*, 2017). Sustainable consumption (SC) is an important way to achieve the goal of sustainable development (Sudarkodi, 2009). Globally, in recent years, there has been an enormous growth in production and consumption, linked with simultaneous environmental degradation. There are many concerns: due to increased consumption, mineral and renewable resources are depleted, and the improvement in economic standards does not necessarily mean improvement in the quality of life. It is well known that a large proportion of the world's population does not have a better quality of life and has been left behind by the industrialised states (Sudarkodi, 2009). Consumers purchase goods to satisfy their needs (food and shelter) and also to have social relationships and a certain social status (Salazar *et al.*, 2013). A multitude of goods are purchased, used and disposed of by consumers (Ahmad *et al.*, 2010), who are aware of the environmental and social aspects, but largely still do not consume with concern (Jonkutė, 2015). A reduction of the consumption volume and changing consumers' habits are the main objectives of SC according to Nagypál *et al.* (2015). Individual behavioural changes should be related to policy measures, such as pricing directed at consumers, informing consumers through product labelling, and utilisation of social marketing campaigns seeking to change the 'unsustainable' habits of individuals (Welch & Southerton, 2019).

In this context, we wanted to investigate consumers' knowledge, attitudes and practices in SC by applying a structured questionnaire. The main questions of this research are:

- Are people aware of the condition of the environment, its pollution, and recycling and reuse of solid waste?
- Do they believe that environmental education is closely linked to poverty?
- What are people's attitudes towards food consumption, recycling and reuse of materials?
- What are the practices of consumers in terms of storing food and avoiding waste?
- Do consumers make efforts to buy recyclable products or avoid consuming resources?

It is important to find the answers to these questions in order to understand how people view SC and whether they are willing to participate in it and, implicitly, in sustainable development, as well as to determine what roles education and poverty play.

Previous studies, published by various authors, were investigated: the importance of planning and shopping routines on food waste generation, in which 244 Romanian consumers were interviewed online by Stefan *et al.* (2013); the environmental knowledge and behaviour of 105 young Romanian pupils (Kolbe, 2014); the opinions and behaviours of the Romanian population with regard to selective collection of solid waste (Pop *et al.*, 2015); and the behaviours, habits and attitudes of 100 young Romanians concerning food waste generation at household level (Ghinea & Ghiuta, 2019). From these studies we know that: the Romanian consumers' intent is to not waste food, but this is not transferred into behaviours due to their planning and shopping routines (Stefan *et al.*, 2013); Romanian students agreed only in part with the waste hierarchy, giving more importance to recycling rather than reuse of waste, and both landfill and incineration were viewed with scepticism by a high proportion of the interviewed students (Kolbe, 2014); 99% of a total of 425 Romanian people knew the meaning of selective collection of waste and 22.6% declared that they did not collect waste separately (Pop *et al.*, 2015); and one of the reasons for food waste is the purchasing of too much food (Ghinea & Ghiuta, 2019). Pop *et al.* (2015) and Omran and Schiopu (2015) state that there is a wide discrepancy between what people claim about recycling, selective collection, and current behaviours regarding solid waste. Instead, Pop *et al.* (2015) affirm that over 60% of the Romanians interviewed were open to increasing their efforts in selective waste collection and beyond. The main factor that influences consumer behaviour is considered to be education (Pop *et al.*, 2015). Many studies prove that people with a higher level of education have a greater perception and awareness of environmental issues (Apaydin & Szczepaniak, 2017; Banyte *et al.*, 2010; D'Souza *et al.*, 2006; Yuan & Zuo, 2013). Apaydin and Szczepaniak (2017) suggest that consumers with a higher level of education and higher incomes are more concerned about environmental protection.

Knowing that poverty and environmental degradation are related, it is important to link poverty and sustainable development (UNEP, 2012). In Romania, the relative poverty rate was 23.6% in 2017 (INS, 2018). Analysis of the poverty rate by gender in 2014–2017 does not reveal significant differences between men and women. However, the risk of poverty affects the population with different intensity depending on the age group, the work capacity available and the income earned (INS, 2018). Throughout the 2014–2017 period, the highest incidence of poverty was among children and young people up to the age of 18, about one-third of whom were below the poverty line, well above the levels corresponding to adults. The state of poverty during this period was maintained at high levels among young people between the ages of 18 and 24. The elderly population (65 years and over) was affected to a lesser extent by poverty (one in five), a level that was lower than the average for the whole population (INS, 2018). Poverty is unevenly distributed even in the regional profile. In 2017, the poverty rate was more than five times higher in the Northeast and Southwest Oltenia regions and more than four times higher in the Southeast and South Muntenia regions, than in the Bucharest–Ilfov region. In 2017, the highest poverty rates were registered in Northeast and Southwest Oltenia (33.4%) and the Southeast (29.6%) regions, and the lowest in Bucharest–Ilfov (6.1%). From the point of view of the relative poverty rate, Romania (with 25.3%) placed in 2016 in the group of relatively poor states with values above the European Union average, including Bulgaria (22.9%), Spain (22.3%), Lithuania (21.9%), Latvia (21.8%), Estonia (21.7%), Greece (21.2%), Italy (20.6%), Croatia (19.5%) and Portugal (19.0%) (INS, 2018).

An investigation of consumers' knowledge, attitudes and practice could provide useful information to improve and encourage sustainable production and consumption. It is helpful to ascertain the level of knowledge of consumers in order to focus on what changes should be made in education and what knowledge should be improved. Also, in order to understand decision-making strategies, it is essential to analyse consumer attitudes, which can only be changed if the attitude is understood. It is well known that knowledge should be correlated with practice, but in reality many consumers do not transfer their knowledge into practice. In this context, the aim of this chapter is to provide some answers regarding the knowledge, attitudes and practice of Romanian consumers. The answers will be validated and compared using the statistical methods applied in this study.

2 Background

Sustainable consumption is 'the use of goods and services that respond to basic needs and bring a better quality of life, while minimizing the use of natural resources, toxic materials and emissions of waste and pollutants over the life cycle, so as not to jeopardize the needs of future generations' (Oslo Symposium, 1994). 'Quality of life', 'environmental protection' and 'ensuring the needs of future generations' are the three important components of SC (Ahamad & Ariffin, 2018). The way in which we consume is one of the fundamental challenges of the present, and everyone contributes to a lesser or greater extent. In this context, it is necessary to understand human behaviour regarding SC. Information can be gathered from particular populations by performing a survey. To find out what a specific population knows, believes and does in terms of SC, a knowledge, attitude and practice (KAP) survey can be developed and applied. The first KAP survey was conducted in the 1950s to understand family planning (Launiala, 2009). Since then, this type of survey has been applied in many different studies: to investigate the level of KAP of women towards waste management (Almasi *et al.*, 2019) or household recycling (Babaei *et al.*, 2015); to evaluate the factors that influence the KAP of local households towards land-use change programmes (Tran *et al.*, 2018) or factors that influence sustainable development (Hai *et al.*, 2010); to examine KAP related to cutaneous leishmaniasis (Sorocco *et al.*, 2017); to investigate seafood safety (Zanin *et al.*, 2015) or food safety (Baş *et al.*, 2006; Kunadu *et al.*, 2016); to assess KAP in SC (Ahamad & Ariffin, 2018) and others. A characteristic of these KAP studies is that they use fewer resources and are more cost-effective than other studies. KAP studies are easy to design, provide quantifiable data and show concise results. The results of these studies tell us what people know, how they feel and how they deal with certain things (Launiala, 2009). The steps included in conducting a KAP survey are (WHO, 2008):

- define the survey objectives;
- develop the survey protocol;
- design the survey questionnaire;
- implement the KAP survey;
- analyse the data;
- use the data.

The basic elements of a KAP survey are domain and target audience identification, sampling method determination (questionnaire), analysis and reporting. The shortcomings of KAP surveys include inaccurate interpretation of data, lack of a standardised approach to validating results, and survey data that may derive from cultural and language differences (Ratcliffe, 1976). The structure of the survey differs from one study to another. Some examples include: 12 items for both knowledge and attitude, 20 items for SC practice (Ahamad & Ariffin, 2018); five items each for environment, economy, society and education (Biasutti & Frate, 2017); 14 items for knowledge, 15 items for attitudes and 12 items for behaviours (Borges, 2019).

There are many studies related to KAP in SC that can be found in the literature (Biasutti & Frate, 2017; Borges, 2019; Ghinea & Ghiuta, 2019; Jonkutė, 2015; Nagypál *et al.*, 2015). All of these studies aim to investigate consumer behaviour. Jonkutė (2015) interviewed 240 Lithuanian respondents in order to investigate their general knowledge and attitudes regarding sustainability. Results showed that 71% highlighted the importance of products and services sustainability. Most of the respondents' choices were determined by limited financial resources, lack of political authority initiatives and distrust of product information (Jonkutė, 2015). Hungarian students were asked to respond to a survey prepared by Nagypál *et al.* (2015) in order to find out their attitudes towards SC. Based on their results, Nagypál *et al.* (2015) stated that sustainable consumption exists in theory, but it should also be evident in practice. Borges (2019) questioned 168 students from a public Portuguese university and 94.6%

agreed with the statement that 'sustainable development entails a reflection on the meaning of quality of life', while 98.8% agreed that 'every girl or boy should receive education that teaches the knowledge, perspectives, values, issues and skills for sustainable living in a community'. Biasutti and Frate (2017) measured the attitudes towards sustainable development of 484 Italian students. Their results showed that psychology students are oriented to social issues, whereas agriculture students showed a higher pro-environmental attitude. Ghinea and Ghiuta (2019) questioned 100 Romanian students in order to discover their behaviours, habits and attitudes regarding food waste. The results showed that one reason for food wastage was consumers purchasing higher volumes of food. 'Expiration date', 'products remaining in the fridge for too long', 'improper storage' or 'products looking bad' were the main contributors to food waste generation (Ghinea & Ghiuta, 2019). The limitations of these studies are: the small size and cultural homogeneity of the group (Biasutti & Frate, 2017; Borges, 2019; Jonkutė, 2015; Nagypál *et al.*, 2015); the influence of the sampling techniques leading to discrepancies; the results being unable to be extrapolated (Stefan *et al.*, 2013); and self-reporting sometimes being biased (Stefan *et al.*, 2013). Various types of software were used for statistical analysis of the results, such as: Statistical Package for Social Sciences (SPSS), version 24 (Borges, 2019); IBM SPSS Statistics 20 and Lisrel 8.80 (Biasutti & Frate, 2017); STATISTICA 12 (StatSoft) and SPSS 17.0 (IBM) (Jonkutė, 2015); and Minitab (Ghinea & Ghiuta, 2019).

The aim of our research is to provide and improve a survey model on SC that has the potential to be applied anywhere in the world, as well as to provide statistical analysis of the survey results with easy-to-use software and their interpretation.

Firstly, a KAP survey was developed in order to investigate SC among Romanian people. The survey was then made available on Google Forms and distributed to different people. Finally, the results were investigated by applying statistical analysis and interpretation.

3 Sustainable Production and Consumption in Romania

Financial growth, the distribution of financial benefits to the entire population in the fight against poverty, ensuring human rights and environmental protection are the main objectives of sustainable development (Dumitru & Sandru, 2015). Changing consumption and production patterns is another important aspect targeted by sustainable development. At the country level, there are short-, medium- and long-term strategic objectives included in the national strategy for sustainable development. The new *Romania's National Sustainable Development Strategy 2030* adopted in 2018 proposed 'a step-by-step transition to a new development model through the introduction of elements of a circular economy, a boosting of resource productivity and a reduction in food loss and waste' (RNSDS, 2018).

In 2018, the population of Romania was 19,530,631 according to INS (2019). The net nominal average monthly earnings per national totalled €519 in 2017, which varies depending on the type of employment. For example, €412 was received by an employee who worked in agriculture, forestry and fishing, €375 by an employee from the food industry and €331 by an employee from the clothing manufacturing industry (INS, 2019). In October 2018, 4,946,800 persons were employed in Romania according to Trading Economics (2019). The living wage for a family in Romania decreased to €480 per month in 2018 from €488 per month in 2017 (Trading Economics, 2019). The Consumer Price Index (CPI) was 100.51 in October 2018 and increased to 103.84 in June 2019. The cost of food increased from 3.73% in October 2018 to 5% in June 2019 based on the data presented by Trading Economics (2019). Also, producer prices increased from 109.1 index points in October 2018 to 114 index points in May 2019 (Trading Economics, 2019).

According to Eurostat (2019a), resource productivity is calculated based on the Gross Domestic Product (GDP) at market prices (reference year 2010) and Domestic Material Consumption (DMC) (index, 2000 = 100) (Figure 10.1). DMC in Romania is very high compared to the average of the EU. Resource Productivity (RP) is the ratio between GDP per capita and DMC per capita. Based on the data obtained from Eurostat (2019a), in 2018 the RP for EU-28 was 2.23 purchasing power standards (PPS)/kg, while in Romania RP was 0.48 PPS/kg.

Compared with the EU average, the RP of Romania is lower, and continued efforts are being made at national level to improve it. Changing mentalities through education, modifying consumer behaviour and developing financial mechanisms would help the transition from the actual model of production and consumption to a circular economy. Reducing food waste during production and consumption is another way to bring Romania closer to sustainable development (RNSDS, 2018).

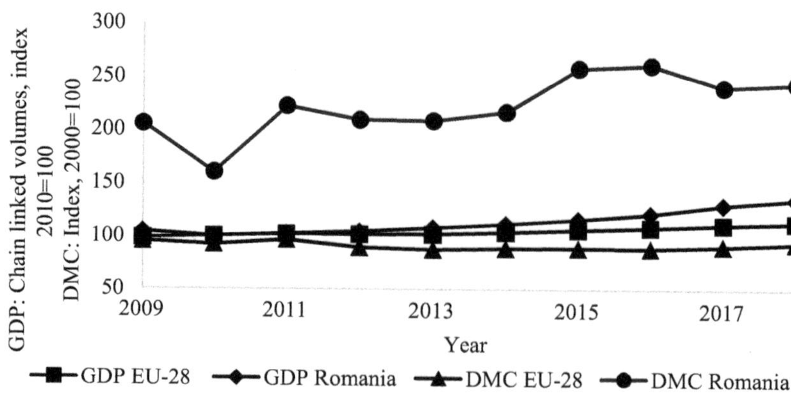

Figure 10.1 GDP and DMC in EU-28 and Romania (2008–2018)

Romania has a transitional temperate continental climate with some influences; the warmest year was 2007 and 1940 was the coldest. The total greenhouse gas (GHG) emissions in Romania in 2011 were 123,359.15 Gg CO_2-eq. (MECC, 2013). The energy sector (69.97%), followed by the agriculture sector (15.35%) and industrial processes (10.22%) are the main contributors to the total amount of GHG emitted. The are several policies and measures for the reduction of GHG emissions such as generation of energy from renewable sources, improvement of buildings' thermal performance, development of ecological houses, utilisation of biofuels, water consumption reduction and use of public transportation (MECC, 2013).

The total amount of water consumed in Romania decreased from 1035.4 million m^3 in 2012 to 740.7 million m^3 in 2016; 76.1% of the total amount of water consumed in Romania is registered to households. Between 2009 and 2017, Romania improved water quality and access to the water supply network, according to RNSDS (2018).

At EU-28 level, the amount of waste generated decreased from 523 kg per capita in 2007 to 486 kg per capita in 2017. In Romania the quantity of waste generated also decreased, from 391 kg per capita in 2007 to 272 kg per capita in 2017 according to Eurostat (2019b). Energy recovery from waste was reported by the EU to be 131 kg per capita while only 12 kg per capita was reported by Romania in 2017. Waste recycling in Romania is a long way behind the EU average. Only 20 kg per capita of waste is recycled in Romania, while the average in the EU is 144 kg per capita, as reported in 2017 (Eurostat, 2019b). Disposal of municipal waste in Romania is predominately to landfill (Schiopu & Ghinea, 2013).

Since both resource use and waste generation continue to increase, a fundamental aim is to reduce the quantity of waste generated and decouple economic growth from environmental degradation (Ghinea *et al.*, 2012, 2017). An important issue is plastic consumption as a large volume of plastic waste is discarded. This type of solid waste can persist in the environment for a long period of time (Comanita *et al.*, 2016). Globally, 230 million tons of plastic were produced in 2009 (Comanita *et al.*, 2015). Among the methods proposed by Comanita *et al.* (2016) for the reduction of wastewater pollution from plastic waste were the recycling of plastics, improvements in the production and consumption of conventional plastic, incineration of plastics for energy recovery and increasing production of bioplastics.

Around 72–76 kg of food waste per capita is generated in Romania (Ghinea & Ghiuta, 2019) and this amount could be composted at composting plants or in households (Ghinea *et al.*, 2019). In Romania, the composting process is underutilised or not utilised at all, but consistent efforts are being made to apply this process (Ghinea *et al.*, 2019).

4 Materials and Methods

We developed a structured questionnaire based on the literature models (Ahamad & Ariffin, 2018; Ahmad *et al.*, 2010; Borges, 2019;

Table 10.1 Items used in the survey

No.	Item	Category
Ki1	The condition of the environment has a direct impact on human health.	Knowledge items (Ki)
Ki2	Ecosystem services are affected by human behaviour.	
Ki3	Human society is largely responsible for climate change.	
Ki4	Do you think that we have enough natural resources (water, fossil fuels, vegetation, etc.) available?	
Ki5	Natural resources should be preserved for future generations.	
Ki6	Do you think that environmental pollution has now reached alarming levels?	
Ki7	Environmental factors (air, water, soil) must be protected against pollution.	
Ki8	The use of plastic should be reduced because it has a negative impact on the environment.	
Ki9	Reuse and recycling should be applied to significantly reduce the amount of waste generated.	
Ki10	Reuse and recycling can reduce the quantity of resources used.	
Ki11	Environmental education is closely linked to poverty.	
Ai1	I care about the condition of the environment because nature gives us our basic needs.	Attitude items (Ai)
Ai2	I appreciate the beauty of the place where I live.	
Ai3	I like products made from recycled materials.	
Ai4	I reuse glass, plastic and paper.	
Ai5	Social development must be a priority in terms of protecting the environment.	
Ai6	More efforts should be made to protect the environment.	
Ai7	The use of renewable energy sources should be increased.	
Ai8	Recycling is difficult to achieve.	
Ai9	I am pleased with the way in which recycling takes place in the country.	
Ai10	Do you think that recycling campaigns are useful?	
Ai11	I use reusable bags.	
Ai12	It is difficult to use reusable bags.	
Ai13	Do you think that too much food is thrown away?	
Ai14	I compare food prices.	
Ai15	I make a shopping list before going to the store.	
Ai16	I tend to buy more than I need.	
Ai17	I was raised with the belief that food should not be wasted.	
Pi1	I feel bad when I throw away unconsumed food.	Practice items (Pi)
Pi2	I think food should not be wasted.	
Pi3	It is difficult for me to store food in the required conditions.	

(*Continued*)

Table 10.1 *(Continued)*

No.	Item	Category
Pi4	It is difficult for me to store certain types of food.	
Pi5	I intend to consume leftover food.	
Pi6	I intend not to throw away food.	
Pi7	I buy products that can be recycled.	
Pi8	I make efforts to avoid consumption of resources.	
Pi9	I use public transport.	

Ghinea & Ghiuta, 2019) to investigate consumers' KAP in SC. Table 10.1 presents the items used in this survey. The questions were selected by considering consumers' behaviours towards SC highlighted by other studies (Ahamad & Ariffin, 2018; Ghinea & Ghiuta, 2019).

The questionnaire was uploaded on Google Forms, posted online and distributed in different social groups. Respondents were chosen based on age and gender. They received a direct (online) invitation to complete the questionnaire. The research was carried out in January–February 2019, during which time 226 people responded to the survey. Only people who had access to the internet and the Google platform were able to answer the questionnaire. This is one of the limiting factors of the research, since the people who responded to the survey had IT skills, and there are many people in the country who do not have these skills or do not have access to the internet. Face-to-face questionnaires will be used in future studies in order to extend the research. In this study, the survey responses were statistically analysed using Minitab 17 software. Item analysis was performed, and the one-way ANOVA method was applied.

5 Results and Discussion

5.1 Interpretation of questionnaire results

Of the total 226 people who answered, 79.2% were female ($n = 179$) and 20.8% were male ($n = 47$). Respondents' ages ranged from 18 to 64 years old (Figure 10.2); 17.7% had been educated to high school level, 73% had university education and 9.3% had undertaken postgraduate studies. Of those interviewed, 60.6% lived in urban areas and 39.4% in rural areas.

Among those who responded to the questionnaire, 43.8% ($n = 99$) said they had incomes below the national average, 34.5% ($n = 78$) had an average salary, 15.9% ($n = 36$) had an above-average salary and 5.8% ($n = 13$) had a much higher salary than the national average.

In order to measure their knowledge related to SC, the respondents were asked to answer 'true' or 'false' to 11 items (Ki). Each correct answer

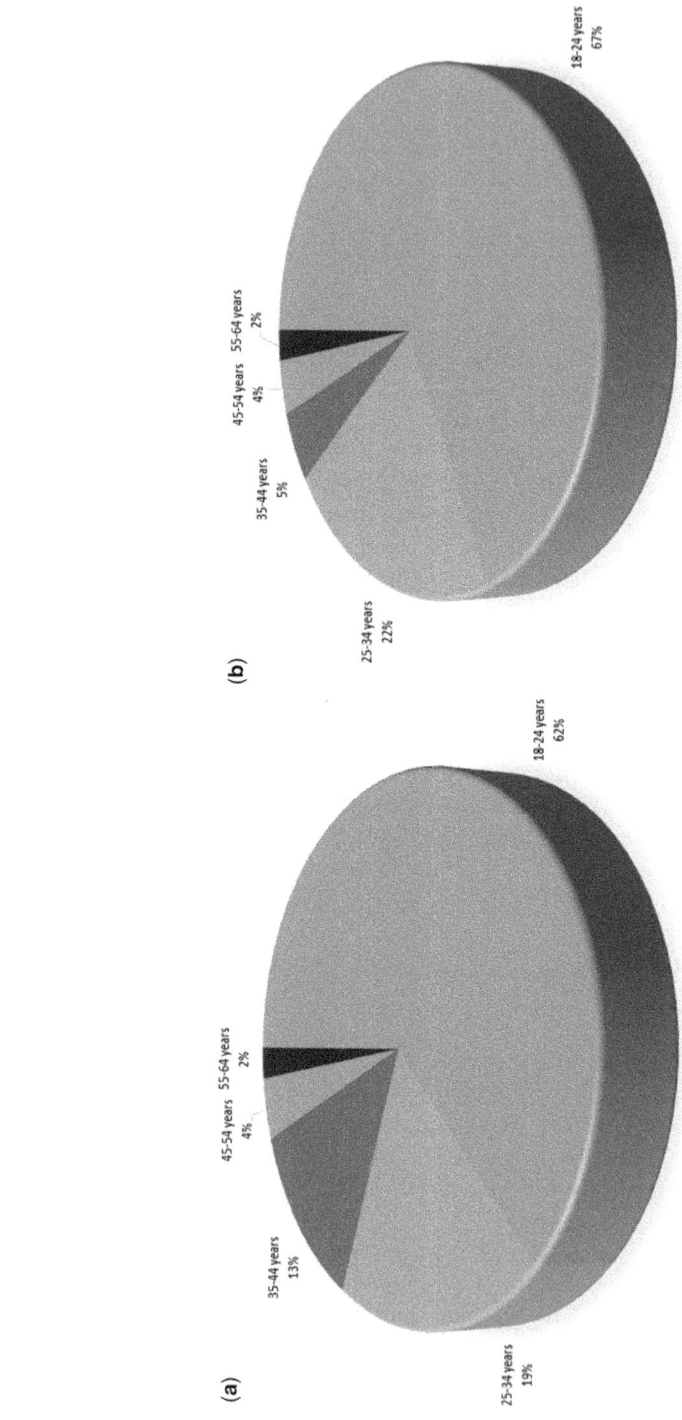

Figure 10.2 Respondents' age: (a) males, (b) females

was given a score of 1, while incorrect answers received a score of 0. The maximum score was 11 and the minimum was 0. Individual results were used to group scores as follows: (i) scores between 80% and 100% represented a high knowledge level, (ii) scores between 60% and 80% demonstrated a moderate level of knowledge and (iii) scores below 60% represented a poor knowledge level. The calculations revealed that the majority of the female respondents, 88.8% ($n = 159$), had a high knowledge level of SC, 9.5% ($n = 17$) had a moderate level of knowledge, while 1.7% ($n = 3$) had a poor level of knowledge. Among the male respondents, 91.48% ($n = 43$) had a high level of knowledge of SC, while 2.12% had a poor knowledge level.

The majority of the female respondents answered correctly to the knowledge items: 99.4% knew that environmental quality has a direct impact on human health and that environmental factors (air, water, soil) must be protected against pollution. Of the female respondents, 177 (98.8%) believed that reuse and recycling should be applied to significantly reduce the amount of waste generated, while 98.3% ($n = 176$) agreed that ecosystem services are affected by human behaviour. Over 90% of female respondents answered correctly to Ki5, Ki6, Ki8 and Ki10. Only 158 (88.2%) female respondents believed that human society is largely responsible for climate change, while 117 (65.3%) believed that there are enough natural resources (water, fossil fuels, vegetation, etc.) still available. The statement that environmental education is closely linked to poverty was agreed to by 44.6% ($n = 80$) of the female respondents. Items Ki1, Ki7, Ki9 and Ki10 gained 100% ($n = 47$) correct answers, showing that male respondents knew that environmental quality is closely related with human health, that reuse and recycling are important ways to reduce waste and that the environment must be protected. Moreover, 91.5% ($n = 43$) of the males who answered this survey agreed that environmental pollution has reached alarming levels, while 8.5% ($n = 4$) disagreed. According to 95.7% ($n = 45$) of the male respondents, natural resources should be preserved for future generations, while 4.3% ($n = 2$) did not agree. Over 97.8% ($n = 46$) of the male respondents agreed that ecosystem services are affected by human behaviour and that the usage of plastic should be reduced. Moreover, 10% ($n = 5$) of the male respondents believed that human society is not largely responsible for climate change, while 38% ($n = 18$) disagreed with the statement that environmental education is closely linked to poverty. Furthermore, 46.8% ($n = 22$) of males responded that there are still enough natural resources (water, fossil fuels, vegetation, etc.) available.

Attitude items (Ai = 17) were developed in order to find out how people feel about SC. For these items, respondents were asked to state whether they agreed, disagreed or were not sure. Each correct answer was scored with 1, wrong answers received –1, while 'not sure' answers were given 0. The attitude level was divided into three categories: a high score (between 80% and 100%), a moderate attitude score (between 60% and 80%) and a poor score (less than 60%). Of the female respondents, 45.26% ($n = 81$) had a moderate score for attitude towards SC, while the percentages of those with high and poor attitude scores were equal at 27.37% ($n = 49$) each. The majority of the male respondents also had a moderate attitude score regarding SC (38.29%), followed closely by those with a poor level (34%), and the remainder demonstrating a high score.

From Table 10.2, it can be observed that a very high percentage of both female and male respondents cared about the condition of the

Table 10.2 Percentages of responses to 'agree' (A), 'not sure' (NS) and 'disagree' (D) of female and male respondents for attitude items

No.	Item	Female			Male		
		A [%]	NS [%]	D [%]	A [%]	NS [%]	D [%]
Ai1	I care about the condition of the environment because nature gives us our basic needs.	96.6	3.4	–	97.9	2.1	–
Ai2	I appreciate the beauty of the place where I live.	95.0	5.0	–	95.7	4.3	–
Ai3	I like products made from recycled materials.	88.8	11.2	–	83.0	17.0	–
Ai4	I reuse glass, plastic and paper.	72.6	27.4	–	61.7	38.3	–
Ai5	Social development must be a priority in terms of protecting the environment.	48.6	15.6	35.8	53.2	12.8	34.0
Ai6	More efforts should be made to protect the environment.	98.3	1.7	–	93.6	4.3	2.1
Ai7	The use of renewable energy sources should be increased.	84.9	12.3	2.8	89.4	6.4	4.3
Ai8	Recycling is difficult to achieve.	26.3	16.8	57.0	14.9	14.9	70.2
Ai9	I am pleased with the way in which recycling takes place in the country.	3.4	5.0	91.6	2.1	12.8	85.1
Ai10	Do you think that recycling campaigns are useful?	79.3	12.8	7.8	70.2	17.0	12.8
Ai11	I use reusable bags.	89.4	10.6	–	87.2	12.8	–
Ai12	It is difficult to use reusable bags.	–	6.7	93.3	–	14.9	85.1
Ai13	Do you think that too much food is thrown away?	95.0	4.5	0.6	91.5	2.1	6.4
Ai14	I compare food prices.	94.4	5.6	–	91.5	8.5	–
Ai15	I make a shopping list before going to the store.	46.9	40.2	12.8	42.6	38.3	19.1
Ai16	I tend to buy more than I need.	58.1	5.0	36.9	44.7	4.3	51.1
Ai17	I was raised with the belief that food should not be wasted.	91.1	8.9	–	89.4	10.6	–

environment, while the remainder were not sure. Over 95% of respondents were aware of and appreciated the beauty of the place where they live. Seventeen percent of male and 11.2% of female respondents were not sure if they like recycled materials, while 72.6% of female and 61.7% of male respondents stated that they reuse glass, paper and plastic. Almost half of respondents of both sexes agreed with the fact that social development must be a priority in terms of protecting the environment and over 90% agreed that more efforts must be made in order to protect the environment. Eighty percent of the respondents felt that the utilisation of renewable energy should be increased, while only a small percentage of the respondents agreed with the way in which recycling takes place in Romania. Over 90% of the respondents believed that food is wasted in large quantities and stated that they compare food prices. Close to 50% confirmed that they buy more than they need and most respondents stated that they were raised with the belief that food should not be wasted.

There were nine SC practice items considered in this study and respondents could choose between the following answers: 'always', 'sometimes', 'rarely' and 'never'. Each answer was given the following score: 1 for 'never', 2 for 'rarely', 3 for 'sometimes' and 4 for 'always'. Based on the percentage of the total individual scores, SC practice levels were ranked as high (scores between 80% and 100%), moderate (scores between 60% and 80%) and poor (scores less than 60%). A moderate practice level was observed in the majority of the female respondents (116, 64.8%) and male respondents (34, 72.3%), while 54 (30.1%) of the female and nine (19.1%) of the male respondents demonstrated a high level of SC practice.

Figure 10.3 illustrates the answers provided by the survey respondents: 49.2% and 34.0% of the female and male respondents, respectively, admitted that they always feel bad about throwing away unconsumed food, and the majority of the respondents agreed that food should not be wasted. Additionally, 35.8% and 34.0% of female and male respondents, respectively, admitted that they sometimes have difficulty in storing food under the required conditions. Over 40% of the respondents answered that they never intend to throw away food, and 67.6% and 48.9% of female and male respondents, respectively, stated that sometimes they buy products that could be later recycled. Also, 65.4% of female respondents answered that they sometimes make efforts to avoid the consumption of resources, while only 48.9% of male respondents gave the same response. In addition, 1.7% of the female respondents stated that they had never used public transport, while the figure was 10.6% for male respondents.

The main cause of environmental degradation was the rapid development of society in the opinion of 68.1% of females and 56.5% of males who responded. However, 23.9% of male and 10% of female respondents strongly believed that environmental degradation is due to poverty, while 19.5% of males and 21.7% of females considered population growth to be the major reason for environmental degradation.

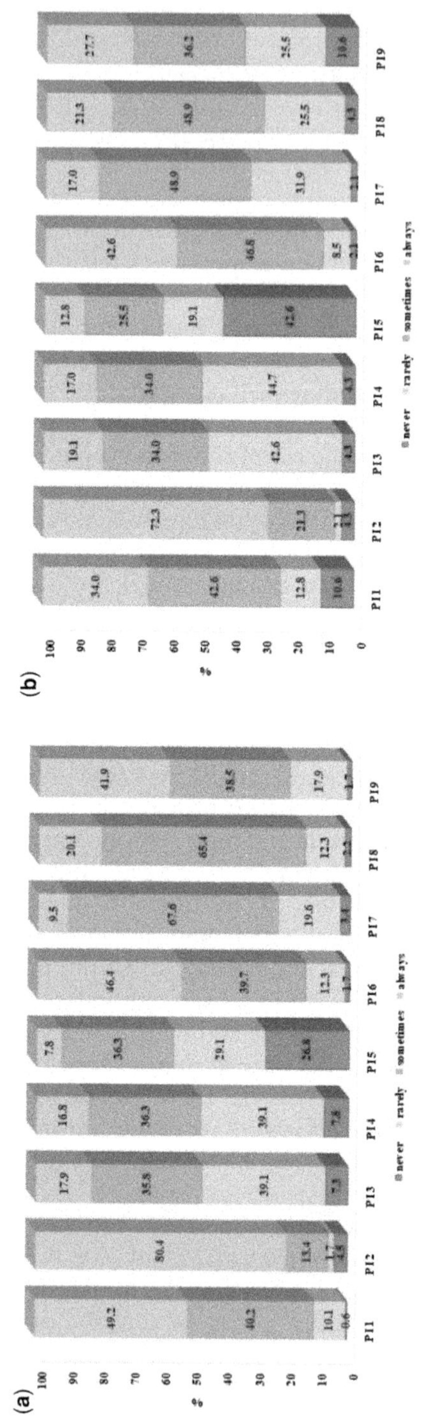

Figure 10.3 Answers of (a) female and (b) male respondents to practice items

5.2 Statistical analysis

5.2.1 Item analysis

For the statistical analysis, all of the items were input into Minitab 17 software. Item analysis was performed in order to establish how a set of items measure KAP related to SC and to find out if there were any problematic questions or affirmations. Table 10.3 presents the knowledge items and total statistics. The total number of females and males who responded to the survey is shown in the column 'Total Count'. The standard deviation (StDev) is a common measure, related to data dispersion. It is thought that lower values for standard deviation indicate lower values for data dispersion. For knowledge items, the values of StDev are between 0.0747 and 0.4986 (female respondents' answers) and from 0.1459 to 0.5044 (male respondents' answers). In order to determine the degree of internal consistency for all included items, Cronbach's alpha was calculated. According to Laerd (2013), Cronbach's alpha values are between 0 and 1, and only values above 0.7 are acceptable. The values obtained for Cronbach's alpha are 0.8044 (female respondents' answers) and 0.8550 (male respondents' answers), meaning that internal consistency is good.

We can use the results presented in Table 10.4 to decide which items can be removed in order to increase the value of the Cronbach's alpha. If Ki11 were removed, the Cronbach's alpha value would increase to 0.8225 (female respondents' answers), while in order to increase the value of this indicator (to 0.8708) for male respondents' answers, the Ki4 item should be removed.

Table 10.3 Knowledge items and total statistics

Variable	Female respondents' answers			Male respondents' answers		
	Total count	Mean	StDev	Total count	Mean	StDev
Ki1	179	0.9944	0.0747	47	0.9787	0.1459
Ki2	179	0.9832	0.1287	47	0.9787	0.1459
Ki3	179	0.8827	0.3227	47	0.8936	0.3117
Ki4	179	0.6536	0.4771	47	0.5319	0.5044
Ki5	179	0.9274	0.2602	47	0.9574	0.204
Ki6	179	0.9553	0.2072	47	0.9149	0.2821
Ki7	179	0.9944	0.0747	47	0.9787	0.1459
Ki8	179	0.9665	0.1805	47	0.9787	0.1459
Ki9	179	0.9888	0.1054	47	0.9787	0.1459
Ki10	179	0.9777	0.1482	47	0.9787	0.1459
Ki11	179	0.4469	0.4986	47	0.617	0.4914
Total	**179**	**9.7709**	**1.7086**	**47**	**9.7872**	**1.9442**

Table 10.4 Omitted knowledge item statistics

Omitted variable	Adj. Total Mean	Adj. Total StDev	Item-Adj. Total Corr	Squared Multiple Corr	Cronbach's alpha
Female respondents' answers					
Ki1	8.777	1.678	0.3932	1.000	0.8037
Ki2	8.788	1.632	0.5714	0.8305	0.7907
Ki3	8.888	1.476	0.6682	0.6389	0.7641
Ki4	9.117	1.395	0.5592	0.5201	0.7933
Ki5	8.844	1.517	0.6968	0.7448	0.7651
Ki6	8.816	1.560	0.6872	0.8131	0.7724
Ki7	8.777	1.678	0.3932	1.000	0.8037
Ki8	8.804	1.583	0.665	0.8276	0.7778
Ki9	8.782	1.653	0.5018	0.7472	0.7967
Ki10	8.793	1.613	0.6149	0.8295	0.7857
Ki11	9.324	1.436	0.4243	0.4282	**0.8225**
Male respondents' answers					
Ki1	8.809	1.837	0.7145	1.000	0.841
Ki2	8.809	1.837	0.7145	1.000	0.841
Ki3	8.894	1.722	0.6669	0.7922	0.8322
Ki4	9.255	1.648	0.4869	0.7053	**0.8708**
Ki5	8.830	1.798	0.6911	0.6519	0.836
Ki6	8.872	1.740	0.6862	0.8178	0.8313
Ki7	8.809	1.837	0.7145	1.000	0.841
Ki8	8.809	1.837	0.7145	1.000	0.841
Ki9	8.809	1.837	0.7145	1.000	0.841
Ki10	8.809	1.837	0.7145	1.000	0.841
Ki11	9.170	1.633	0.5436	0.7246	0.8609

Note: Higher values obtained for Cronbach's alpha are in **bold**.

The obtained values for StDev for attitude items are between 0.129 and 0.954 (female respondents' answers) and from 0.146 to 0.987 (male respondents' answers). For the attitude items, the Cronbach's alpha was 0.9243 (female respondents' answers) and 0.9432 (male respondents' answers). Both values are above 0.9 which means that internal consistency is excellent. Using the results presented in Table 10.5, we can decide which attitude items can be removed in order to increase the value of the Cronbach's alpha. If Ai16 were removed, the Cronbach's alpha value would increase to 0.928 (female respondents' answers), while in order to increase the value of this indicator (to 0.947) for male respondents' answers, the Ai5 item should be removed.

Table 10.5 Omitted attitude item statistics

Omitted variable	Adj. Total Mean	Adj. Total StDev	Item-Adj. Total Corr	Squared Multiple Corr	Cronbach's alpha
Female respondents' answers					
Ai1	10.899	5.751	0.565	0.923	0.924
Ai2	10.916	5.709	0.657	0.959	0.922
Ai3	10.978	5.607	0.776	0.952	0.919
Ai4	11.140	5.507	0.764	0.879	0.917
Ai5	11.994	5.197	0.680	0.942	0.923
Ai6	10.883	5.801	0.411	0.605	0.926
Ai7	11.045	5.484	0.809	0.876	0.916
Ai8	11.559	5.145	0.798	0.921	0.917
Ai9	10.983	5.542	0.741	0.977	0.918
Ai10	11.151	5.334	0.852	0.904	0.913
Ai11	10.972	5.613	0.773	0.956	0.919
Ai12	10.933	5.674	0.711	0.895	0.921
Ai13	10.922	5.691	0.636	0.942	0.922
Ai14	10.922	5.697	0.678	0.929	0.922
Ai15	11.525	5.226	0.893	0.906	0.911
Ai16	12.078	5.226	0.609	0.906	**0.928**
Ai17	10.955	5.636	0.757	0.949	0.920
Male respondents' answers					
Ai1	10.617	6.691	0.437	1.000	0.945
Ai2	10.638	6.635	0.583	1.000	0.944
Ai3	10.766	6.434	0.837	0.898	0.938
Ai4	10.979	6.378	0.754	0.834	0.939
Ai5	11.787	6.157	0.604	0.859	**0.947**
Ai6	10.681	6.521	0.653	1.000	0.941
Ai7	10.745	6.384	0.785	1.000	0.938
Ai8	11.043	6.075	0.901	1.000	0.935
Ai9	10.766	6.383	0.850	1.000	0.937
Ai10	11.021	6.099	0.910	1.000	0.934
Ai11	10.723	6.473	0.830	1.000	0.939
Ai12	10.745	6.449	0.845	1.000	0.939
Ai13	10.745	6.374	0.731	1.000	0.939
Ai14	10.681	6.541	0.751	1.000	0.941
Ai15	11.362	6.095	0.857	0.914	0.936
Ai16	11.532	6.036	0.691	0.860	0.945
Ai17	10.702	6.504	0.799	1.000	0.940

Note: Higher values obtained for Cronbach's alpha are in **bold**.

The StDev values for practice items were determined, with values between 0.621 and 0.941 (female respondents' answers) and from 0.707 to 1.1 (male respondents' answers). The practice items obtained the following values for Cronbach's alpha: 0.9688 (female respondents' answers) and 0.9710 (male respondents' answers). Based on our statistics for omitting individual practice items, we could say that the value of the Cronbach's alpha could increase from 0.968 to 0.971 (female respondents' answers) if the Pi2 item were not considered for the analysis. It was also observed that the same practice item (Pi2) could be omitted in order to increase the value of Cronbach's alpha from 0.971 to 0.972 (male respondents' answers).

5.2.2 One-way ANOVA

Based on *p*-value determination, with the one-way ANOVA method, we can establish whether the differences between group means are statistically significant. The group means can be examined by using an interval plot and the Tukey procedure for comparison. Calculation of S, R^2 and R^2 pred. could provide insight into how well the model fits the data (Ghinea & Ghiuta, 2019). The answers for KAP items were evaluated. In this study the following parameters were considered: null hypothesis = all means are equal, alternative hypothesis = at least one mean is different, significance level $\alpha = 0.05$ (the most commonly used value) (Ghinea & Ghiuta, 2019; Pennsylvania State University, 2017). According to Ghinea and Ghiuta (2019), 'if p-value $\leq \alpha$ the null hypothesis is rejected and the alternative hypothesis is considered; if p-value $> \alpha$ the null hypothesis is not rejected'.

(a) One-way ANOVA – knowledge items:

- Not all group means are equal since the *p*-values for female and male respondents' answers are $0.000 < \alpha$.
- Ki11 has the lowest mean (0.4469) in the case of female respondents' answers, and Ki1 and Ki7 have the highest mean (0.9944) (Figure 10.4a). Ki4 has the lowest mean (0.5319) in male respondents' answers and six knowledge items (Ki1, Ki2, Ki7, Ki8, Ki9 and Ki10) have the highest mean (0.9787) (Figure 10.4b). The difference between the mean of Ki11 and other items' means is probably significant because the interval bars do not overlap (Figure 10.4a), while the interval bars for Ki1, Ki2, Ki7, Ki8, Ki9, Ki10 means overlap and it is obvious that there is no difference between these means (Figure 10.4b). The pooled standard deviation is 0.2670 in the case of female respondents' answers and 0.2766 for male respondents' answers.
- The Tukey method and 95% confidence were used to compare the group means. The results showed that for the female respondents' answers, Ki1, Ki2, Ki5, Ki6, Ki7, Ki8, Ki9 and Ki10 are included in Group A, and Ki5, Ki6 and Ki8 are also included in Group B. Ki3 is

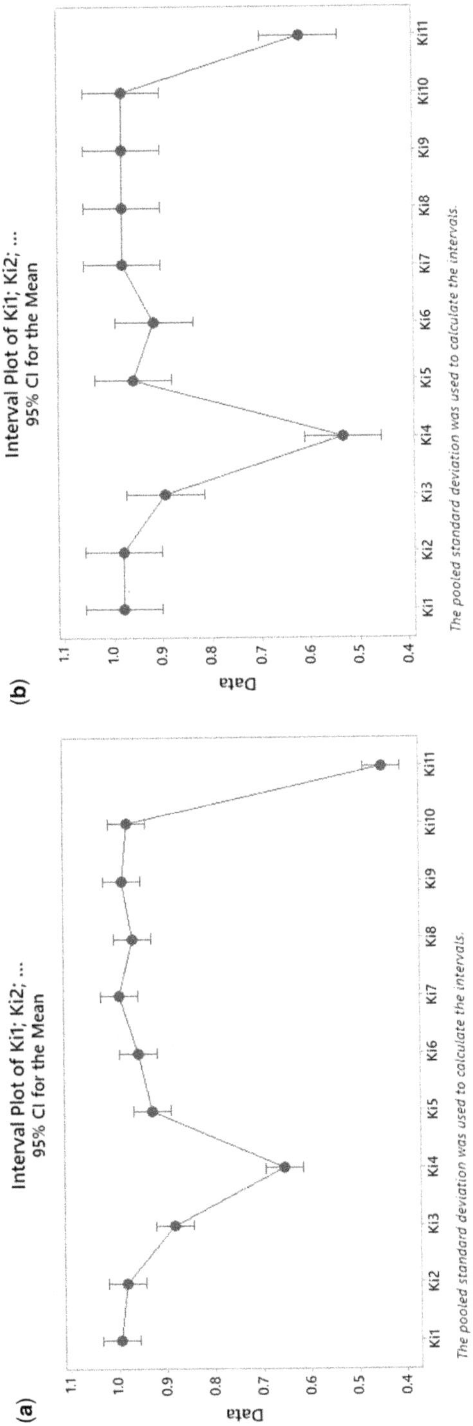

Figure 10.4 Interval plot for (a) female and (b) male respondents' answers for knowledge items

only in Group B, while Ki4 and Ki11 are in two different groups, C and D. The items that share a group are not significantly different, while the others that do not share a group are significantly different. In the case of the male respondents' answers, it was observed that Ki4 and Ki11 are in the same group (B), while the other knowledge items are in Group A.

- S and R^2 were calculated in order to see how well the model fits the data. For the female respondents' answers to the knowledge items, the following values were obtained: $S = 0.267$, $R^2 = 28.55\%$, R^2 (pred.) = 27.74%, which means that the model does not predict new observation. It can be seen that the model explains only 28.55% of the variation in the response. In the case of the male respondents' answers, the values were: $S = 0.276$, $R^2 = 23.65\%$, R^2 (pred.) = 20.29%. These values are lower than the previous but are strongly related to the number of male respondents.

(b) One-way ANOVA – attitude items:

- The p-values are equal with 0.000 for female and male respondents' answers, and we can say that not all group means are equal, since p-values < α.
- The highest mean in the case of female respondents' answers for attitude items was observed for A16 (0.983), while the lowest mean was registered for Ai8 (0.307). Two different attitude items have the highest value for mean (Ai1, 0.978) and the lowest value (A16, 0.064) in male respondents' answers. For responses from female respondents, the pooled standard deviation is 0.5121, while for male respondents this is 0.5491.
- After applying the Tukey method, the results indicated that the female respondents' answers Ai1, Ai2, Ai3, Ai6, Ai7, Ai9, Ai11, Ai12, Ai13, Ai14, Ai17 are included in Group A and Ai3, Ai7, Ai9, Ai11 are included in all three groups, A, B and C. Ai17 is in A and B groups and Ai4 is in B and C groups. Ai10 is only in Group C, while Ai8 and Ai15 are in Group D and Ai5 and Ai16 in Group E. In the case of male respondents' answers, it was observed that Ai1 is included only in Group A and Ai2 is included in both Groups A and B. Groupings for Ai3, Ai4, Ai6, Ai7, Ai9, Ai11, Ai12, Ai13, Ai14, Ai17 are not significantly different and they are included in three groups, A, B and C. Ai10 is also in three groups, B, C and D, while Ai8 is in C and D. Ai15 is not significantly different from Ai10, Ai8 and Ai16, being included in Groups D and E. Ai16 is in Groups E and F, while Ai5 is only in F.
- The model does not predict new observation: $S = 0.512$, $R^2 = 34.69\%$, R^2 (pred.) = 33.95% for female respondents' answers and $S = 0.549$, $R^2 = 25.57\%$, R^2 (pred.) = 23.85% for male respondents' answers.

(c) One-way ANOVA – practice items:

- We can say that not all group means are equal since p-values < α (p-value = 0.000 for both female and male answers regarding practice items).
- Pi5 has the lowest mean (2.251) in the case of female respondents' answers for practice items and Pi2 the highest (3.698). With a value of 2.085, Pi5 has the lowest mean and Pi2 the highest (3.617) also in the case of male respondents' answers. The pooled standard deviations are 0.764 and 0.852, respectively.
- Female and male respondents' answers for practice items were evaluated by applying the Tukey method, and it was observed that Pi5 is significantly different from other items since this item does not share a letter group with the others.
- As in the case of the other items, it was observed that the model does not predict new observation: $S = 0.764$, $R^2 = 24.66\%$, R^2 (pred.) = 23.81% and $S = 0.852$, $R^2 = 18.78\%$, R^2 (pred.) = 15,21%.

6 Brief Summary

Two hundred and twenty-six Romanian people answered a questionnaire about knowledge, attitudes and practices in SC. The questions included in the survey were selected based on other studies and translated into the Romanian language. The main findings from this study are:

- the majority of the female (88.8%) and male (91.48%) respondents had a high level of knowledge of SC. Also, it was observed that both female (45.26%) and male (38.29%) respondents had moderate attitude levels regarding SC, while a moderate practice level was observed for 64.8% of female and 72.3% of male respondents, respectively;
- almost all of the people who participated in this survey knew that environmental quality influences human health. Most of them thought that ecosystem services are affected by human behaviour and they agreed that more efforts should be made in order to protect the environment;
- some of the respondents strongly believed that environmental degradation is due to poverty and that population growth is one major reason for environmental degradation;
- item analysis and the one-way ANOVA can be successfully applied for statistical analysis of the survey answers.

7 Discussion

The type of questionnaire used in this study has a three-dimensional structure, allowing measurement of KAP related to SC. People who answered this survey had a high level of education, with most of them having undertaken university education. This is also observed in the answers given by these people, hence the high level of knowledge regarding SC. The high level of knowledge regarding SC was also reported by Borges (2019), Jonkutė (2015) and Nagypál *et al.* (2015). In our study more than half of the respondents lived in the city, so it is difficult to extrapolate results for the rural population. In future studies we will apply the questionnaire to a larger number of people living in both urban and rural areas. A deficiency of the study is the distribution of the questionnaire only in an online environment, to a select group of people (who have a computer, who know how to work in an online environment, who have a certain education and a certain income to be able to afford a computer and access to the internet). However, most of the respondents had lower or average incomes. Another minus would be that people can be subjective, even if the name of the person completing the questionnaire was not requested. As it was completed online, it was not possible to explain any potential misunderstandings. In contrast, an online questionnaire has benefits such as its low cost and requirement for minimal resources, a short time frame for completing and collecting data, and the assurance that those who complete it have a certain level of education. However, in future studies the survey will be conducted face-to-face, in order to explain possible uncertainties and to take into account other groups of people. The time required for data collection, as well as processing and interpreting results, in this case, will be longer. The results of this study show that there is a difference between the answers given by females and those given by male respondents regarding the claim that environmental education is related to poverty: 44.6% of females agreed with this statement while 38% of males disagreed. Almost all survey respondents agreed that more efforts are needed to protect the environment. The results showed that both women and men who responded to this survey had a moderate attitude and practice level regarding SC. This aspect (high knowledge level and moderate/low attitude or practice scores) was also observed by Pop *et al.* (2015), Omran and Schiopu (2015) and Nagypál *et al.* (2015). The statistical analysis of the data obtained from the survey consisted of analysis of items with calculation of Cronbach's alpha values, as in the studies of Borges (2019), Ghinea and Ghiuta (2019), Apaydin and Szczepaniak (2017) and Jonkutė (2015). In these studies, as mentioned above, the software that was used differs. The answers to the items included in the questionnaire were also analysed using one-way ANOVA, the method applied by Apaydin and Szczepaniak (2017), Borges (2019) and Ghinea and Ghiuta (2019).

The results of this study make known the level of KAP of a specific group of Romanian respondents at a given time and can help decision makers in taking measures to make the transition from knowledge to practice possible. The questionnaire developed for this study has the potential to be applied as it is or improved based on the specific cultures and education in other parts of the world.

8 Conclusion

The notions of sustainability and environmental protection must be considered every day, as the Earth has limited resources that should be preserved for future generations as well. The findings of this study indicate a high level of knowledge about sustainable consumption among Romanian respondents, but they show a moderate attitude and practice level. Therefore, it is important to move from theory to practice. This is an issue that should be investigated more in future studies, in order to establish the barriers to converting knowledge about sustainable consumption into practice and to offer solutions to overcome these obstacles. Communication, power of example, education and poverty eradication are the keys to sustainable consumption and, implicitly, sustainable development.

References

Ahamad, N.R. and Ariffin, M. (2018) Assessment of knowledge: Attitude and practice towards sustainable consumption among university students in Selangor, Malaysia. *Sustainable Production and Consumption* 16, 88–98.

Ahmad, S.N., Juhdi, N. and Awadz, S. (2010) Examination of environmental knowledge and perceived pro-environmental behaviour among students of university Tun Abdul Razak, Malaysia. *International Journal of Multidisciplinary Thought* 1, 328–342.

Almasi, A., Mohammadi, M., Azizi, A., Berizi, Z., Shamsi, K., Shahbazi, A. and Mosavi, S.A. (2019) Assessing the knowledge, attitude and practice of the Kermanshahi women towards reducing, recycling and reusing of municipal solid waste. *Resources, Conservation & Recycling* 141, 329–338.

Apaydin, F. and Szczepaniak, M. (2017) Analyzing the profile and purchase intentions of green consumers in Poland. *Ekonomika* 96, 93–112.

Babaei, A.A., Alavi, N., Goudarzi, G., Teymouri, P., Ahmadi, K. and Rafiee, M. (2015) Household recycling knowledge, attitudes and practices towards solid waste management. *Resources, Conservation & Recycling* 102, 94–100.

Banyte, J., Brazionienė, L. and Gadeikiene, A. (2010) Investigation of green consumer profile: A case of Lithuanian market of eco-friendly food products. *Economics and Management* 15, 374–383.

Baş, M., Ersun, A.S. and Kıvanç, G. (2006) The evaluation of food hygiene knowledge, attitudes, and practices of food handlers' in food businesses in Turkey. *Food Control* 17, 317–322.

Biasutti, M. and Frate, S. (2017) A validity and reliability study of the Attitudes Toward Sustainable Development scale. *Environmental Education Research* 23, 214–230.

Borges, F. (2019) Knowledge, attitudes and behaviours concerning sustainable development: A study among prospective elementary teachers. *Higher Education Studies* 9, 22–32.

Comanita, E.D., Ghinea, C., Hlihor, R.M., *et al.* (2015) Challenges and opportunities in green plastics: An assessment using the ELECTRE decision-aid method. *Environmental Engineering and Management Journal* 14, 689–702.

Comanita, E.D., Hlihor, R.M., Ghinea, C. and Gavrilescu, M. (2016) Occurrence of plastic waste in the environment: Ecological and health risks. *Environmental Engineering and Management Journal* 15, 675–685.

D'Souza, C., Taghian, M., Lamb, P. and Peretiatko, R. (2006) Green decisions: Demographics and consumer understanding of environmental labels. *International Journal of Consumer Studies* 31, 371–376.

Dumitru, C.G. and Sandru, A.M. (2015) Sustainable strategic directions of regional development in Romania in the context of European integration. CES Working Paper No. VII (pp. 452–46).

Eurostat (2019a) *Resource Productivity Statistics.* See https://ec.europa.eu/eurostat/statistics-explained/index.php?title=Resource_productivity_statistics#Resource_productivity_of_the_EU-27_and_across_Member_States_over_time (accessed 31 May 2019).

Eurostat (2019b) *Municipal Waste Statistics.* See https://ec.europa.eu/eurostat/statistics-explained/index.php?title=Municipal_waste_statistics (accessed 31 May 2019).

Ghinea, C. and Ghiuta, A.O. (2019) Household food waste generation: Young consumers' behaviour, habits and attitudes. *International Journal of Environmental Science and Technology* 16, 2185–2200.

Ghinea, C., Petraru, M., Bressers, H. and Gavrilescu, M. (2012) Environmental evaluation of waste management scenarios – significance of the boundaries. *Journal of Environmental Engineering and Landscape Management* 20, 76–85.

Ghinea, C., Campean, T. and Gavrilescu, M. (2017) Integrating sustainability indicators for tracking anthropogenic pressure on the earth – the footprint family. *Environmental Engineering and Management Journal* 16, 935–948.

Ghinea, C., Apostol, L.C., Prisacaru, A.E. and Leahu, A. (2019) Development of a model for food waste composting. *Environmental Science and Pollution Research* 26, 4056–4069.

Hai, P.H., Dung, T.A. and Hens, L. (2010) Influencing factors on sustainable development: A case study in Quang Tri province, Vietnam. *Environment, Development and Sustainability* 12, 103–116.

INS (2018) *Dimensions of Social Inclusion in Romania in 2017.* Bucharest: National Institute of Statistics.

INS (2019) *Statistics.* See http://statistici.insse.ro:8077/tempo-online/#/pages/tables/insse-table(accessed 21 January 2019).

Jonkutė, G. (2015) The consumers' approach to sustainable consumption and production: A case study in Lithuania. *Environmental Research, Engineering and Management* 71, 28–46.

Kolbe, K. (2014) Knowledge, attitudes and behaviour regarding waste management options in Romania: Results from a school questionnaire. *Journal of Economic Development, Environment and People* 34, 95–107.

Kunadu, A.P.H., Ofosu, D.B., Aboagye, E. and Debrah, K.T. (2016) Food safety knowledge, attitudes and self-reported practices of food handlers in institutional foodservice in Accra, Ghana. *Food Control* 69, 324–330.

Laerd (2013) *Cronbach's Alpha Using Minitab.* See https://statistics.laerd.com/minitab-tutorials/cronbachs-alpha-using-minitab.php (accessed 21 January 2019).

Launiala, A. (2009) How much can a KAP survey tell us about people's knowledge, attitudes and practices? Some observations from medical anthropology research on malaria in pregnancy in Malawi. *Anthropology Matters Journal* 1, 1–13.

MECC (2013) *Romania's Sixth National Communication on Climate Change and First Biennial Report.* Bucharest: Ministry of Environment and Climate Change. See https://unfccc.int/sites/default/files/6th_nccc_and_1st_br_of_romania%5B1%5D.pdf (accessed 21 January 2019).

Nagypál, N.C., Görög, G., Harazin, P. and Baranyi, R.P. (2015) 'Future generations' and sustainable consumption. *Economics and Sociology* 8, 207–224.

Omran, A. and Schiopu, A.M. (2015) Reasons for non-participation in recycling of solid waste in northern Malaysia: A case study. *Environmental Engineering and Management Journal* 14, 233–243.

Oslo Symposium (1994) *Oslo Roundtable on Sustainable Consumption and Production.* See http://enb.iisd.org/consume/oslo004.html (accessed 10 October 2019).

Pennsylvania State University (2017) *STAT 200. Elementary Statistics. Lesson 10: One-way Analysis of Variance (ANOVA).* See https://online.stat.psu.edu/stat200/lesson/10.

Pop, I.N., Baciu, C., Bican-Brişan, N., Muntean, O.-L., Costin, D. and Rogozan, G.C. (2015) Evaluating attitudes and behaviour towards selective collection of waste in Cluj-Napoca city, Romania. *Studia UBB Ambientum*, LX: 87–101.

Ratcliffe, J. (1976) Analyst biases in KAP surveys: A cross-cultural comparison. *Studies in Family Planning* 7, 322–330.

RNSDS (2018) *Romania's National Sustainable Development Strategy 2030, adopted by the Romanian Government on 9 November 2018 through Government Decision 877/2018.* See http://dezvoltaredurabila.gov.ro/web/wp-content/uploads/2019/03/Romanias-Sustainable-Development-Strategy-2030.pdf (accessed 21 January 2019).

Salazar, H.A., Oerlemans, L. and van Stroe-Biezen, S. (2013) Social influence on sustainable consumption: Evidence from a behavioural experiment. *International Journal of Consumer Studies* 37, 172–180.

Schiopu, A.M. and Ghinea, C. (2013) Municipal solid waste management and treatment of effluents resulting from their landfilling. *Environmental Engineering and Management Journal* 12, 1699–1719.

Sorocco, T., Weigel, M., Armijos, R., Cevallos, W., Sanchez, X. and Puebla, E. (2017) Cutaneous Leishmaniasis knowledge, attitudes and practices (KAP) survey of an endemic rainforest population in northern Ecuador. *Annals of Global Health* 83, 165.

Stefan, V., Van Herpen, E., Tudoran, A.A. and Lähteenmäki, L. (2013) Avoiding food waste by Romanian consumers: The importance of planning and shopping routines. *Food Quality and Preference* 28, 375–381.

Sudarkodi, K. (2009) Achieving sustainable consumption for sustainable development: Issues and solutions. MPRA Paper No. 15755. University Library of Munich.

Tkalec, M.R. (2010) The role of ethics in the education for sustainable development of man. *Objavljeni Znanstveni Prispevek Na Konferenci (1.08)*, BV 70 (4), 577–582.

Trading Economics (2019) *Statistics Romania.* See https://tradingeconomics.com/romania/ (accessed 21 January 2019).

Tran, H., Nguyen, Q. and Kervyn, M. (2018) Factors influencing people's knowledge, attitude, and practice in land use dynamics: A case study in Ca Mau province in the Mekong delta, Vietnam. *Land Use Policy* 72, 227–238.

UNEP (2012) *Sustainable Consumption and Production for Poverty Eradication.* Nairobi: United Nations Environment Programme.

Welch, D. and Southerton, D. (2019) After Paris: Transitions for sustainable consumption. *Sustainability: Science, Practice and Policy* 15, 31–44.

Wesarat, P., Sharif, M.Y. and Majid, A.H.A. (2017) Role of organizational ethics in sustainable development: A conceptual framework. *International Journal of Sustainable Future for Human Security* 5, 67–76.

WHO (2008) *A Guide to Developing Knowledge, Attitude and Practice Surveys.* Geneva: World Health Organization.

Yuan, X. and Zuo, J. (2013) A critical assessment of the Higher Education for Sustainable Development from students' perspectives – a Chinese study. *Journal of Cleaner Production* 48, 108–115.

Zanin, L.M., da Cunha, D.T., Stedefeldt, E. and Capriles, V.D. (2015) Seafood safety: Knowledge, attitudes, self-reported practices and risk perceptions of seafood workers. *Food Research International* 67, 19–24.

11 Environmental Sustainability Concerns: The Case of the Sundarbans of Bangladesh

Azizul Hassan

Environmental sustainability is a concept that has received considerable attention from the relevant researchers. However, environmental sustainability has not been well researched in the context of tourism, creating a research gap in this area. In its simple meaning, environmental sustainability refers to maintaining the factors and practices that contribute to the environment on a longstanding basis. This research aims to explain environmental sustainability in relation to conscience and sustainable business. Consumers of tourism products and services are not equally concerned about the environment across the world. The Sundarbans in Bangladesh is a relevant example. This forest experienced an unexpected oil spill in its recent history. The incident not only damaged the flora and fauna of the forest but also resulted in serious consequences for environmentally sustainable business. This chapter presents conceptual research based on a review of the relevant studies in the literature and online resources. The findings show that environmental sustainability, conscience and sustainable business are interrelated. These concepts can have a significant positive effect on a forest with a fragile and pristine natural setting. The research suggests that the involvement of both national and international agencies plays an effective role in ensuring and promoting environmental sustainability in a given context. This research is thus a contribution towards the limited knowledge in the areas of environmental sustainability, conscience and sustainable business. It concludes that environmental sustainability is a possibility but partly depends on the good practices of business enterprises as well as the positive actions of the relevant parties involved.

1 Introduction

Environmental sustainability is a topic of significance at present, and relates to the understanding of the impacts of human activities on the environment (Howes *et al.*, 2017). Furthermore, environmental sustainability as a subject is gradually receiving recognition from diverse parties, particularly including general beneficiaries, academics, governmental or non-governmental agencies and the media (Hassan *et al.*, 2020a). The growing concern and increased attention are leading to research to assess the impacts of human activities on nature and the environment. This conceptual paper explains environmental sustainability through the lenses of conscience and sustainable business. The context of this research aligns with the example of Bangladesh, a South Asian country. The national tourism policy in Bangladesh arguably offers better scope for environmentally sustainable business (Hassan & Burns, 2014; Hassan & Kokkranikal, 2018; Hassan *et al.*, 2020b). The country recently struggled with an unexpected oil spill in the Sundarbans, the world's largest mangrove forest. This research uses the example of the Sundarbans of Bangladesh and the recent incident. Environmental sustainability is a rather well-researched academic area. However, there is limited knowledge about its relationship with conscience and sustainable business initiatives. This research is thus a contribution to this gap and, with a relevant example, critically explains environmental sustainability.

2 Environmental Sustainability

Environmental sustainability can be defined as the rates of the creation of pollution, the harvesting of renewable resources and the depletion of non-renewable resources that can be continued indefinitely. If these are discontinued for an indefinite time period, they cannot be sustainable (Suhi *et al.*, 2019). Environmental sustainability is viewed as a typical concern that consumers have started to take into consideration. The basic reason for this consideration is education about sustainability; those who are concerned about environmental sustainability are mostly better educated (Schaltegger *et al.*, 2012). A higher education level impacts the attitudes and actions of consumers who, in turn, take actions to develop caring and responsible attitudes towards tourism businesses and the environment.

Addressing sustainability is important. A good number of business model researchers opine that, in order to contribute to sustainable development, organisations need to reconsider their business models (Abdelkafi & Täuscher, 2016). The origin of the notions of sustainability and sustainable development dates back to the late 1980s. Many researchers from diverse disciplines have, since then, outlined these concepts to clarify their meaning and usage and concentrate on defining their weaknesses and strengths (Glavic & Lukman, 2007). In 1987, sustainable development was explained as 'development that meets the needs of the present without compromising the ability of future generations to meet their own needs' (WCED, 1987: 5). The current and future generations' demands are normally explained with reference to the 'triple bottom line', which covers economic, environmental and social factors (Elkington, 1997). This explanation is adopted widely either explicitly or implicitly by many business organisations as it presents the concept of sustainability in a comparatively clear and simple way (Joyce & Paquin, 2016). Sustainability is also observed as conducting change and innovation to cope with the environment within which business organisations operate (Faber *et al.*, 2005).

With both environmental and socioeconomic factors attached to business models, innovation remains relevant in business models aiming to ensure sustainability. In the literature, Business Model Innovation (BMI) is featured as a way in which sustainability can be incorporated into business organisations (Boons *et al.*, 2013; Geissdoerfer *et al.*, 2016; Roome & Louche, 2015). A relevant example of this is the categorisation offered by Cavalcante *et al.* (2011). The categorisation identifies four types of change as relevant to business models: the creation of a business model (conceptualisation of a new business model); the extension of a business model (expansion of a business by maintaining the main processes of the existing one); the revision of the business model (replacement of the business model with a new one); and the termination of a business model (relinquishing the existing business model). This dynamic context can be explained with reference to two types of sustainability: weak and strong. The strong form of sustainability applies to radical change that concentrates on the whole integration of sustainability in the core business process. The weak form of sustainability refers to incremental changes followed by the adoption of some activities leading to partial transformation and aiming to achieve a sustainable business model (Abdelkafi & Täuscher, 2016). Researchers argue that the incremental changes in the weak form of sustainability are inadequate to contribute to sustainable development. The transformation of businesses thus needs to move further (Boons *et al.*, 2013). Along with environmental sustainability, the transformation of tourism businesses thus also needs to incorporate conscience.

3 Conscience

In many cases, the conscience of a consumer affects their buying behaviour directly. Honey (2008) argues for 'responsible travel to natural areas which conserves the environment and improves the welfare of the local people' (Honey, 2008: 6). The term 'conscience' is generally meant as a person's capacity to differentiate both right and wrong answers on the basis of their ethical or moral values (Ubani, 2012). This notion is very often attached to consumer perspectives. Consumers of tourism services and products are commonly defined as relatively more conscious and aware of their behaviour and responsibilities towards the environment (Boniface, 2001). Thus, it is essential that the demands of these consumers are fully understood, as well as their moral and ethical considerations. The impact of consumer perspectives on sustainability can be seen when a particular product is distinguished as ethical. In the tourism business, in particular, it is important to recognise the influence of consumers' viewpoints on their purchasing attitudes. The conscience of consumers is a reasonable and valid factor contributing to the thought process when selecting and purchasing the most desired product or service.

Consumer conscience is influenced by a number of distinguishable factors that include the role of media, popular opinion, personal values and the government's involvement in ensuring sustainability practices (Ubani, 2012). In earlier times, people commonly ignored attitudes of responsibility towards the environment and nature. Many of the attitudes failed to comply with the environmental needs to ensure pollution-free and sustainable resource use. Thus, the negative impacts of such irresponsible attitudes led to threats to environmental safeguards. A precise understanding of the long-term implications of human attitudes on the environment requires further explanation. Effective policy planning and implementation generally offer competitive advantages to both the locals and the nation as a whole (Honey, 2008). The responsibility for and attention to the environment need to be translated into effective planning and actions.

Feelings often shape and influence a consumer's buying behaviour and decision making. This becomes particularly true when general consumers actively begin to look for products or services that make them feel comfortable about their own demands and environmental well-being (Boniface, 2001). If there is any suggestion of harm to the well-being and development of a surrounding environment, it needs to be given proper attention so as to be either eliminated or reduced (Ubani, 2012). Consumers generally possess the capacity to differentiate tourism products or services that are environmentally friendly or unfriendly. As expected, consumers are more likely to purchase sustainable tourism products or services that are environmentally friendly. It is very unlikely that an entrepreneur or marketer would wish to see their products or services disregarded by consumers. To prevent this possibility, tourism products or services need to be environmentally friendly in order to allow consumers to get what they intend to purchase (Boniface, 2001). This is important because a lack of concern for the environment and conscience by tourism businesses can lead to a disaster that harms sustainability.

4 A Disaster for Sustainability: The Sundarbans Example

The Sundarbans (UNESCO World Heritage Site) is the world's largest mangrove and tidal halophytic mangrove forest, covering a region of

10,000 km^2 in both Bangladesh and India (Rahman & Hassan, 2016). The Sundarbans, as the habitat of many different species of animals, has very often been disturbed by human activities. Fortunately, in recent years there have been few such disturbances, and with less severity. However, the oil spill was described as a major incident. According to the *Daily Prothom Alo* (2015), on 9 December 2014 an oil spill took place in the Shela River in the Sundarbans. The site of the incident is near the Mongla Port of Bangladesh and about 100 km away from the Kolkata Port of India. An oil tanker named the 'Southern Star VIII', carrying 350,000 litres of furnace oil, was in a collision with a cargo vessel. The oil tanker was anchored at the confluence of the Passur River and the Shela River. The cargo vessel collided with the tanker in dense fog at 5:00 am. The oil tanker sank in the river, and by 17 December the oil had spread over an area of 350 km^2 in canals and another river, blackening the nearby shoreline. Oil spills seriously threaten the natural habitats of both flora and fauna, including marine resources with dolphins, fish, plankton, trees and forest-dependent peoples (AFP, 2014). The incident took place in a protected mangrove area that is considered the home of the Ganges and Irrawaddy dolphins.

The oil spill seriously threatened the natural habitat and surrounding populations as environmentalists warned that the incident was an ecological 'catastrophe', for the forest itself and for the dolphins in particular. The gross loss in the Sundarbans was uncountable. Experts also raised concern over the negative effects on the well-being of aquatic organisms.

According to *National Geographic* (2010), the 10 species put at risk by an oil spill are the Ganges and Irrawaddy dolphin, royal Bengal tiger, great egret, rhesus macaque, black-capped kingfisher, leopard, northern river terrapin, chital, saltwater crocodile and horseshoe crab. The oil spill evidently caused the death of a few Irrawady dolphins. The thick layer of oil on the river threatened the lives both in the natural habitat and also among the surrounding populations that tend to rely on the river. It hampered the aquatic organisms and weeds. Wildlife near the river is at constant risk of death because of the potential for mishaps, and the oil layer acted as a barrier to food collection. Salinity is the other issue in terms of sustainable tourism business in the Sundarbans (World Bank, 2017a). This makes evident that conscience, sustainability and tourism business are symmetrical and deserve adequate attention.

5 Conscience, Sustainability and Tourism Enterprises

Tourism can possibly be viewed as a factor of development in a developing country like Bangladesh. However, if tourism is not properly regulated, it can lead to great sociocultural, economic and environmental damage for the communities living in the areas concerned (Apollo, 2015). Ecotourism is an alternative type of tourism, developed as a panacea for the situation in developing countries across the world. A considerable number of definitions of ecotourism are offered in relevant studies in the literature. Many researchers consider ecotourism to be a means through which countries can create sustainable development. This can be seen in examples set by international organisations like the United Nations Environment Programme (UNEP). This organisation includes ecotourism as an element of their agendas of sustainability. The relationship of ecotourism to sustainable development is far from linear and is a much debated topic among researchers and scholars (UNEP, 2015).

Sustainability both as a practice and as a procedure does not necessarily have a negative impact on a business's profit-maximisation capacities (Bocken *et al.*, 2014). The actual benefits from sustainability practices to a tourism business are cost reduction and the enhancement of competitiveness (UNWTO, 2017). Thus, it is obvious that environmental sustainability practices in tourism are a possibility with the effective participation of tourism business operators and beneficiaries. Each of the measures for environmental sustainability can be capitalised by both return of investment and bottom-line improvement. The most crucial benefits that sustainability practices in tourism businesses can offer are the acceleration of business functionalities and competitiveness with good brand image in a global market setting.

According to Sheffi and Blanco (2018), sustainable business initiatives are demanded in countries that have fragile and sensitive natural settings. Tourism has been responsible for the development of certain business initiatives and many of these are featured as sustainable. By creating sustainable efforts as part of a business initiative or enterprise, that particular business entity makes a commitment for sustainability to become an inseparable element of the bottom line. In this way, any business decisions, partnerships and supply chains are all featured as sustainability initiatives. Sustainable business initiatives can hardly create a threat to the environmental and natural setting of a tourist attraction. Thus, acceptability also increases across the world. In tourism, sustainable business concepts are important in the sense that they are friendlier to the environment and the natural setting. Thus, a sustainable tourism business practice is deemed socially responsible, economically viable and environmentally friendly. This is normally viewed as being sustainable. Tourism businesses need to have policies that are socially responsible, economical, friendly and supportive for overall growth and development. These enterprises also deserve to have environmental sustainability and values that align with consumer conscience.

6 Environmental Sustainability and Conscience in the Sundarbans

In spite of excellent progress and development, a good number of the local communities around the Sundarbans live below the poverty line

(World Bank, 2014). This is one reason why the World Bank describes the Sundarbans as both an inspiration and a challenge for policymakers and practitioners of development (World Bank, 2017b). Thus, it is evident that tourism in the Sundarbans has experienced a poor image and has not been planned and implemented properly. This actually validates one of the key reasons why tourism in the Sundarbans needs to be emphasised in a sustainable way (Khanom *et al.*, 2004).

The potential for tourism in the Sundarbans is encouraging and business activities surrounding tourism are also encouraging and highly rated (Hassan, 2012; Hassan & Sharma, 2017a, 2017b). International NGO Relief International, supported by EU funding, created eight eco-sites in the Sundarbans in 2013. The aim is to encourage local communities to incorporate tourism enterprises. Relief International is a non-profit humanitarian agency providing development assistance across the world. This specific programme in Bangladesh involves the local communities and grassroots movements for dealing with issues of health, poverty and the environment (Relief International, 2016). The project has gone some way towards enabling the local communities to benefit from tourism. It has reduced their dependency on the resources of the Sundarbans. Relief International has provided support for the generation of supplementary income through entrepreneurship (Relief International, 2016).

For decades, Bangladesh has been struggling to ensure environmental sustainability through tourism development in the Sundarbans. Due to limited access, transportation arrangements and a lack of facilities, particular sufficient ammodation, tourist number remains low (United Nations Educational, Scientific and Cultural Organization, 2019). This number of tourists gradually decreased in later years, suffering from some major drawbacks including a lack of sustainable tourism approaches.

Thus, it is evident that tourism in the Sundarbans has suffered from a poor image and has not yet taken off in a real sense. The potential for ecotourism is encouraging and business activities surrounding ecotourism have also been encouraging and highly rated. These actually encompass one of the many diverse routes within which the communities living around the Sundarbans can evolve and grow (Khanom *et al.*, 2004).

7 Conclusion

The aim of this chapter was to understand environmental sustainability through conscience and sustainable business. Both notions are interrelated in the process of ensuring environmental sustainability. In practice, environmental sustainability receives attention from the concerned parties as it is an area that remains an important consideration for tourism businesses. The Sundarbans of Bangladesh is a good example of a region that has experienced serious environmental consequences, given the recent oil spill. The research explained and analysed business models that address environmental sustainability. A critical explanation outlined the ways in which a disaster affects environmental sustainability and tourism business, from the perspective of the oil spill in the Sundarbans Natural World Heritage Site (WHS). Sustainable business practices became challenging due to this drastic incident. However, it is encouraging that people's awareness of and commitment to the reduction of pollution and the protection of nature is increasing, with the Sundarbans as an example. Such practices may lead to more positive attitudes towards the environment and the support of environmental sustainability. From a global market standpoint and based on key findings, the research thus suggests that it is necessary to clearly understand the influences of any disaster on environmental sustainability and tourism business in particular. In addition, the findings show that a disaster leaves a negative effect on environmental sustainability, resulting in the shift of badly affected tourism businesses to other, better areas. This research is conceptual and relied on a review of the available studies in the literature. This is rather a limitation. Research with primary data input may improve understanding of the basic issues and opportunities. Future similar research studies are expected to include primary data.

References

Abdelkafi, N. and Täuscher, K. (2016) Business models for sustainability from a system dynamics perspective. *Organization & Environment* 29, 74–96.

AFP (2014) Bangladesh oil spill 'threatens rare dolphins'. *The Guardian*, 11 December. See http://www.theguardian.com/environment/2014/dec/11/bangladesh-oil-spill-threatens-rare-dolphins (accessed 1 March 2019).

Apollo, M. (2015) The clash – social, environmental and economical changes in tourism destination areas caused by tourism. The case of Himalayan villages (India and Nepal). *Current Issues of Tourism Research* 5 (1), 6–19.

Bocken, N., Short, S., Rana, P. and Evans, S.A. (2014) Literature and practice review to develop sustainable business model archetypes. *Journal of Cleaner Production* 65, 42–56.

Boniface, P. (2001) *Dynamic Tourism: Journeying with Change*. Clevedon: Channel View Publications.

Boons, F., Montalvo, C., Quist, J. and Wagner, M. (2013) Sustainable innovation, business models and economic performance: An overview. *Journal of Cleaner Production* 45, 1–8.

Cavalcante, S., Kesting, P. and Ulhøi, J. (2011) Business model dynamics and innovation: (Re-)establishing the missing linkages. *Management Decision* 19, 1327–1342.

Daily Prothom Alo (2015) *Long-term Damage to Sundarbans Feared*. See https://en.prothomalo.com/bangladesh/Long-term-damage-to-Sundarbans-feared (accessed 1 March 2019).

Elkington, J. (1997) *Cannibals with forks – Triple Bottom Line of 21st Century Business*. Stoney Creek, CT: New Society Publishers.

Faber, N., Jorna, R. and van Engelen, M.L. (2005) The sustainability of 'sustainability' – a study into the conceptual foundations of the notion of 'sustainability'. *Journal of Environmental Assessment Policy and Management* 7 (1), 1–33.

Geissdoerfer, M., Bocken, N. and Hultink, E. (2016) Design thinking to enhance the sustainable business modelling process – a workshop based on a value mapping process. *Journal of Cleaner Production* 135, 1218–1232.

Glavic, P. and Lukman, R. (2007) Review of sustainability terms and their definitions. *Journal of Cleaner Production* 15, 1875–1885.

Hassan, A. (2012) Package eco-tour as special interest tourism product – Bangladesh perspective. *Developing Country Studies* 2 (1), 1–8.

Hassan, A. and Burns, P. (2014) Tourism policies of Bangladesh – a contextual analysis. *Tourism Planning and Development* 11 (4), 463–466.

Hassan, A. and Kokkranikal, J. (2018) Tourism policy planning in Bangladesh: Background and some steps forward. *e-Review of Tourism Research (eRTR)* 15 (1), 79–87.

Hassan, A. and Sharma, A. (2017a) Wildlife tourism for visitor learning experiences: Some evidences on the Royal Bengal Tiger in Bangladesh and India. In I. Lima and R. Green (eds) *Wildlife Tourism, Environmental Learning and Ethical Encounters* (pp. 155–168). New York: Springer.

Hassan, A. and Sharma, A. (2017b) Wildlife tourism: Technology adoption for marketing and conservation. In J.K. Fatima (ed.) *Wilderness of Wildlife Tourism* (pp. 61–85). Waretown, NJ: Apple Academic Press.

Hassan, A. Umachandran, K. and Halim, H.S. (2020a) Environmental sustainability on tourism business enterprise: Model and case. In A. Sharma, J.I. Pulido and A. Hassan (eds) *Sustainable Destination Branding and Marketing: Strategies for Tourism Development* (pp. 154–162). Wallingford: CABI.

Hassan, A., Kennell, J. and Chaperon, S. (2020b) Rhetoric and reality in Bangladesh: Elite stakeholder perceptions of the implementation of tourism policy. *Tourism Recreation Research* 45 (3), 307–322.

Honey, M. (2008) *Ecotourism and Sustainable Development: Who Owns Paradise?* Washington, DC: Island Press.

Howes, M., Wortley, L., Potts, R., Dedekorkut-Howes, A., Serrao-Neumann, S., Davidson, J., Smith, T. and Nunn, P. (2017) Environmental sustainability: A case of policy implementation failure? *Sustainability* 9 (165), 1–17.

Joyce, A. and Paquin, R.L. (2016) The triple layered business model canvas: A tool to design more sustainable business models. *Journal of Cleaner Production* 135, 1474–1486.

Khanom, S., Shah, M. and Chaudhary, A. (2004) Towards ecotourism: Issues of current tourism practices in the Sundarban mangrove forest, Bangladesh. Paper presented at the Peace, Environment and Tourism Conference, Pokhara, Nepal.

National Geographic (2010) Marine habitat destruction, coastal areas are bearing the brunt. *National Geographic*, 27 April. See http://ocean.nationalgeographic.com/ocean/critical-issues-marine-habitat-destruction/ (accessed 1 March 2019).

Rahman, M. and Hassan, A. (2016) Sustainable tourism practices and development: The case of the Sundarbans. *Tourism Dimensions* 3 (1), 17–22.

Relief International (2016) *Linking Jute to Jobs: Ecotourism in Bangladesh*. See https://www.ri.org/from-the-field/linking-jute-jobs-ecotourism-bangladesh (accessed 1 March 2019).

Roome, N. and Louche, C. (2015) Journeying toward business models for sustainability: A conceptual model found inside the Black Box of organizational transformation. *Organization & Environment* 29, 11–35.

Schaltegger, S., Lüdeke-Freund, F. and Hansen, E.G. (2012) Business cases for sustainability: The role of business model innovation for corporate sustainability. *International Journal of Innovation and Sustainable Development* 6 (2), 95–119.

Sheffi, Y. and Blanco, E. (2018) *Balancing Green: When to Embrace Sustainability in a Business (and When Not To)*. Cambridge, MA: Massachusetts Institute of Technology.

Suhi, S.A., Enayet, R., Haque, T., Ali, S.M., Moktadir, M.K. and Paul, S.K. (2019) Environmental sustainability assessment in supply chain: An emerging economy context. *Environmental Impact Assessment Review* 79, 106306.

Ubani, L.U. (2012) *Hospitality and Tourism Human Relations Management in Africa*. Bloomington, IN: AuthorHouse.

UNEP (United Nations Environment Programme) (2015) *UNEP Year Book 2014: Emerging Issues in our Global Environment.* New York: United Nations Publications.
United Nations Educational, Scientific and Cultural Organization (2019) *The Sundarbans.* See https://whc.unesco.org/en/list/798/ (accessed 1 March 2019).
UNWTO (United Nations World Tourism Organisation) (2017) *Sustained Growth in International Tourism Despite Challenges.* See https://www.unwto.org/archive/global/press-release/2017-01-17/sustained-growth-international-tourism-despite-challenges.
WCED (World Commission on Environmental Development) (1987) *Our Common Future: Report of the World Commission on Environment and Development.* Oxford: Oxford University Press.
World Bank (2014) *A Resilient Future for Bangladesh Sundarbans.* See http://www.worldbank.org/en/news/feature/2014/07/02/a-resilient-future-for-bangladesh-sundarbans (accessed 1 March 2019).
World Bank (2017a) *Increasing Salinity in a Changing Climate Likely to Alter Sundarban's Ecosystem.* See http://www.worldbank.org/en/news/feature/2017/01/22/increasing-salinity-in-a-changing-climate-likely-to-alter-sundarbans-ecosystem (accessed 1 March 2019).
World Bank (2017b) *Where We Work – Bangladesh Overview.* See http://www.worldbank.org/en/country/bangladesh/overview (accessed 1 March 2019).

12 Urbanisation, Access to Affordable Housing and Sustainable Development among Low-Income Earners in Nigeria: Insights from the Bariga Local Council Development Area, Lagos State

Kennedy Eborka, Chinyere C. Nwokeke and Mosope Akinbola

Although urbanisation is often seen as an indication of positive development, for many developing countries the process of urbanisation tends to produce negative consequences in the form of lack of access to liveable housing, the emergence of slums and the high housing cost burden among low-income urban dwellers. When low-income earners (LIEs) in urban areas spend a large part of their income on housing-related issues, their ability to save and make future investments in education, health and capital accumulation is compromised, thereby threatening the sustainability of their future. This study was carried out in the Bariga Local Council Development Area of Lagos, Nigeria. The probability sampling technique was used to select 160 respondents. The aim of the study was to explore the living conditions of LIEs in the Bariga community. Data generated were subjected mainly to descriptive analysis. Major findings from the study showed that the majority of the respondents (34.37%) earned 20,000 naira or less per month, while only 10% earned 50,000 naira or more. Currently, naira (₦), the Nigerian currency, exchanges at ₦1 to US$0.0028. This means that the majority of the respondents earned about

US$56 per month. About 45% lived in one-room apartments, and 57.1% of the respondents spent over 30% of their monthly income on housing-related issues. Among other things, the study recommends that governments should shift from the usual elitist approach to the provision of housing and place greater emphasis on housing for low- and medium-income earners. Unless efforts are increased in this direction, the attainment of Sustainable Development Goal (SDG) 1, especially, the focus on building the resilience of the poor, will be difficult to achieve among LIEs in urban centres of Nigeria.

1 Introduction

In the last few decades, the growing population, together with the effects of globalisation and rural–urban migration, have seen many developing countries, including Nigeria, become urbanised (PRB, 2018). The process of urbanisation is sometimes viewed as an indication of societal transformation that comes with improvements in the human condition (Beall *et al.*, 2010). Thus, when a society is said to be urbanised or has a high level of urbanisation, it creates a picture of an appreciable standard of living among its inhabitants (Duranton, 2009). The question is whether rapid urbanisation really brings about improvement in the living conditions of the typical urban dweller. Far from promoting widespread improvements, urban areas are usually a mosaic of people with different socio-economic characteristics and backgrounds that affect their abilities to cope with the costs and demands of city life (Badru, 2004; Tettey, 2007).

The aim of this study was to examine housing affordability among low-income residents in the Bariga community, Lagos State. Low-income earners in urban centres go through severe stress in trying to meet their basic needs with a low income against the backdrop of high levels of unemployment, inflation and costs of living. The study concentrated on investigating the types of housing available to LIEs in Bariga, and explored the housing cost burden and the proportion of income that goes towards the housing needs of low-income residents. Understanding the housing needs of LIEs is important for policymakers and development planners in view of the global aim to eradicate poverty in all its forms as encapsulated in the declarations of the Sustainable Development Goals (SDGs).

2 Background

Nigeria is among the fastest urbanising countries in the Global South. In 2017, the country's level of urbanisation was 49%; this increased to 50% in 2018 (PRB, 2017, 2018). If this increase persists, it is estimated that by 2028, 60% of the country's over 200 million people will be living in urban centres (PRB, 2019). Although the process of urbanisation tends to be associated with development (Beall *et al.*, 2010), the literature indicates that rapid urbanisation poses serious challenges for the provision of adequate and affordable housing among LIEs in urban areas (Chidebell, 2013).

The ability of this population, especially those in the Global South, to acquire other basic human needs remains a big problem. Many countries in the Global South are plagued with low purchasing power parity (PPP) which affects citizens' abilities to maintain a decent living. PPP in most African countries is weak, with a few exceptions such as Algeria (US$15,050), Mauritius (US$22,570) and South Africa (US$13,090) (PRB, 2018). For instance, the figure for Nigeria is US$5680, while Angola, Ghana and Uganda have figures of, respectively, US$6060, US$4490 and US$1820. When these amounts are compared with those of the Global North and some parts of Asia, the difficulties in accessing liveable housing among LIEs in African countries becomes clearer. To illustrate this, the figures for Singapore, the UK, the United States, China, Poland and Italy are, respectively, US$90,570, US$43,160, US$60,200, US$16,760, US$28,170 and US$40,030 (PRB, 2018). The implication of low PPP is weak purchasing power to access basic human necessities such as housing and enhanced quality of life (Habibi & Zebardast, 2018). Housing has been regarded as the most important predictor of life satisfaction, and a liveable shelter constitutes one of the essential human needs that are vital for physical, emotional and cognitive survival and quality of life. As Adugna (2006) argues, in the evaluation of a person's comfort, growth and development, housing remains a critical element.

Given the high rate of population growth, rural–urban migration, unemployment and underemployment in Nigeria, together with the existing minimum wage of ₦18,000 per month (about US$50.40), a substantial number of residents in urban centres in Nigeria will be unable to afford safe and decent housing. Consequently, the development of slums and squatter settlements and their associated negative consequences, such as increased criminality and cultural and environmental degradation, will become inevitable (Fahmi, 2014).

In all events, affordable decent housing is a major problem for the urban poor, and many who cannot afford to pay live in the streets and open spaces, in some cases with their children (Chidebell, 2013). Those who can afford accommodation pay a significant part of their incomes as rent, leaving little or nothing for savings and investments in education and skill acquisition. This situation is likely to affect the attainment of sustainable cities and communities, as well as the eradication of poverty in all its forms, both of which are important targets on the SDG agenda. Affordable housing has been defined as a form of housing that does not cost more than 30% of the income of the occupants or household (McNamara, 2008; UN, 2002). Therefore, any household that spends more than 30% of its income on housing is under a severe burden.

In Nigeria, while successive governments have over the years embarked on several housing intervention programmes with the objective of making housing available and affordable to the majority of the population, LIEs are barely remembered by city authorities in the course of urban development. In many cases, urban development plans are driven by a capitalistic push that tends towards clearing cities of their lower class without regard for their social and economic well-being. The manner of governments' interventions in the housing sector is elitist, making affordable housing for many LIEs a desire that is impossible to attain. Thus, housing problems, particularly shortages and affordability, persist. It was this concern that prompted the impetus for this study. The study was guided by the following research questions: What types of housing are available to LIEs in Bariga? What proportion of income goes towards the housing needs of low-income residents? To what extent does this affect their ability to save for future investment?

3 Materials and Method

This study was conducted in the Bariga Local Council Development Area (LCDA) of Lagos State, Nigeria. Lagos State is one of the six states in the South-West geopolitical zone in Nigeria. The state is the most urbanised in the country and serves as the nation's economic hub, making it a preferred destination for migrants from other parts of the country. The population of the state is estimated to be about 20 million people (NPC, 2006), and is thus one of the most populous states in Nigeria. Bariga is a suburb in Lagos State and was formerly under the Shomolu Local Government Area (LGA). In 2003, the community was made an LCDA with headquarters in Gbagada, a neighbouring community. The study adopted the non-experimental research design under which a cross-sectional survey was utilised for data gathering. The probability sampling technique was used to select 160 respondents. The selection of respondents was made in stages. First, the streets in the community were listed. Next, the simple random probability sampling technique was adopted to select 11 streets from those listed. Occupied compounds/buildings were recorded from the selected streets and the systematic sampling technique was used to select some buildings. Bariga is a low-income area with numerous cases of one building housing several households. To choose a household, the number of households in each selected building was determined, and one household was selected through the simple random sampling technique. An eligible respondent was then selected from the household. An eligible respondent was a man or woman who was the head of the household or who could represent the head of the household and give information on income, level of education, age and amount paid as rent.

The research instrument was a questionnaire with 62 questionnaire items. The instrument was pre-tested, after which the observed inadequacies were addressed before the actual study commenced. The structured interview method of questionnaire administration was adopted and the survey yielded a 100% return rate. Data analysis was mainly descriptive, involving frequencies and percentages and presented in charts to illustrate the findings.

4 Findings

The analysis of data gathered from the field was mainly descriptive, involving univariate analysis. Data were distributed through frequencies and percentages in charts. The sociodemographic characteristics of the participants, such as gender, age group, marital status, education level and employment, among other univariate variables, were captured.

Figure 12.1 is a line graph showing the sex of the respondents. According to the information in the graph, males constituted the majority of the research participants. The distribution indicates that 53.75% of the respondents were males, while 46.25% were females. This could be attributed to the fact that more males than females were available when eligible respondents were selected during the survey. The preponderance of males should therefore not be construed to be the result of disinterest from women. Another possible suggestion is that the availability of more men at home during the survey could be the result of the high rate of unemployment in Nigeria which keeps many men at home (although some may be involved in night shift work), while their wives or spouses go out during the day to engage in petty trading along the streets or to hawk goods on the highways to augment the family income. Nigeria is a strong patriarchal society and many men without a job would find it challenging to engage in commercial activities that are socially constructed as belonging in the domain of women.

Data in Figure 12.2 show that those within the age group 30–39 years constituted the majority of the respondents in the study. Furthermore, Figure 12.3, which presents the marital status of respondents, indicates that about 21% were single, while those that were married constituted 62%. Separated or divorced respondents formed, respectively, 11% and

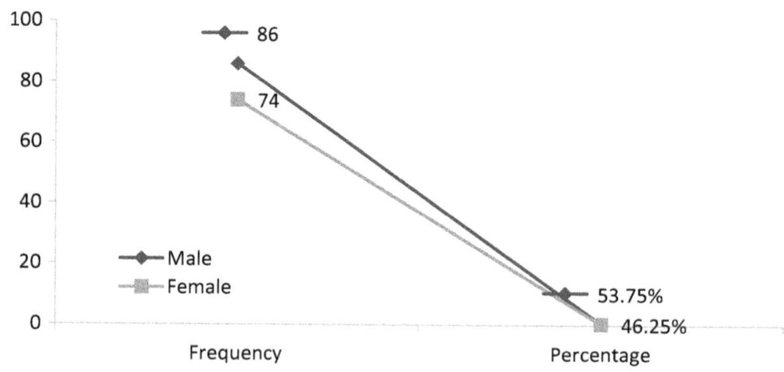

Figure 12.1 Gender of respondents

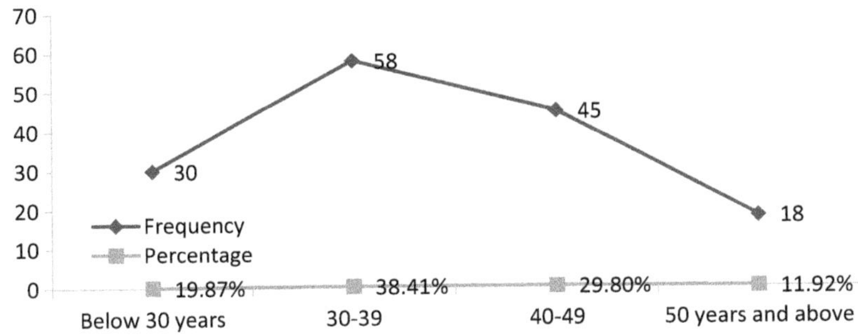

Figure 12.2 Age of respondents

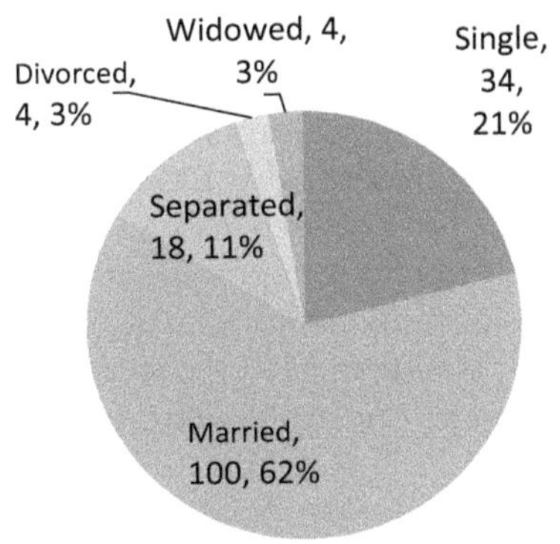

Figure 12.3 Marital status of respondents

3%. This means that the majority of the respondents were in marital unions at the time of the survey. However, the statistics of those in marital unions should be treated with the understanding that a significant number of marriages in Nigerian society are without legal status. Although an increasing number of people go through the process of legalising their conjugal relationships, a great many consider themselves married when they are living together despite any legal recognition. Overall, the data reinforced the observation made in previous studies that marriage is almost universal in Nigerian society (Oyekanmi, 2004). People are expected to marry when they reach a certain age. A grown male or female who is still single is often reminded of this through the words of prayers and wishes directed at him/her in daily conversations. Sometimes this takes on some form of pressure from friends, family and community members.

Figure 12.4 presents data on the ethnic affiliation of respondents. It shows that people from the Igbo ethnic group constituted the majority. This is a little curious when one considers that Lagos State is a South-Western state and home to the Yoruba people. However, it is understandable when one accounts for the fact that Lagos is Nigeria's primate city

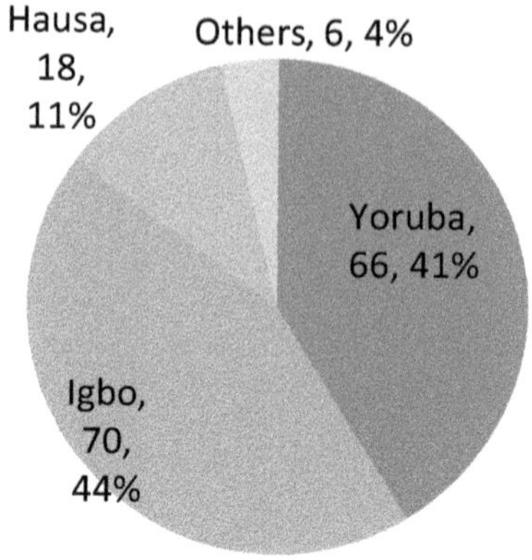

Figure 12.4 Ethnic affiliation of respondents

and the preferred destination of migrants. Nigeria is made up of about 250 ethnic groups, among which the Igbo have been noted to be the most widely travelled and found in almost all states of the Nigerian federation. Additionally, the process of chain migration can inform the phenomenon of people from the same ethnic origin clustering in certain areas. This could explain why the Igbo constituted the majority of respondents in the study. The area captured in the study could be an area with a predominantly Igbo population.

Sixty per cent of the respondents were Christians, while 33% were Muslims. These two religions are the most dominant and pronounced in Nigerian society. Religion is a powerful force in African societies in general, and Nigeria in particular. Many rely on religion to cope with the challenges of existence and the interpretation of certain social and natural phenomena. This has manifested in the surge of religious organisations of different types in the country (Akanle, 2010).

The proportion of respondents with primary level education was 45%; it was 42.50% for those with secondary level and 12.50% for those with tertiary education. The data suggest that all of the research participants had, at least, primary school education. This could be an indication of the effectiveness of the federal government policy on Universal Basic Education. However, the fact that the majority (45%) of the respondents did not obtain education beyond primary level calls for concern. This corroborates an earlier report on education at the national level which indicated a low level of higher educational attainment among the Nigerian population (NPC & ICF International, 2014). To that end, there is a need to intensify efforts at all levels of government (national, state and local) to stimulate an increase in enrolment in higher education. A situation in which the majority of people in a society did not receive an education beyond primary school poses a significant handicap to the drive towards national development and sociocultural transformation.

In terms of the employment status of the participants, the majority were employed at the time of the study. A large proportion of these were self-employed and involved in different crafts and trades. Lagos is a commercial city, and many people survive by engaging in commercial activity to cope with the economic pressure in the state. Figure 12.5 shows the distribution of the nature of employment: 35% of the respondents worked in or were employed in the public sector, 13% were in the private sector and 46% were self-employed, while the 'not applicable' category made up only 6%. The high percentage of those in self-employment helps account for the high level of employment among the respondents. In fact, the majority of working people in Nigeria are self-employed individuals.

Analysis of Figure 12.6 reveals that 34.37% of the respondents earned below ₦20,000 per month, 23.13% made between ₦20,000 and ₦29,999, 15.63% earned between ₦30,000 and ₦39,999 and 16.87% earned

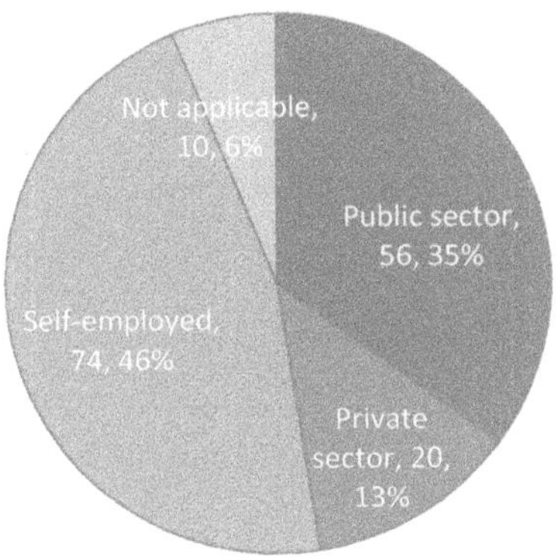

Figure 12.5 Nature of employment

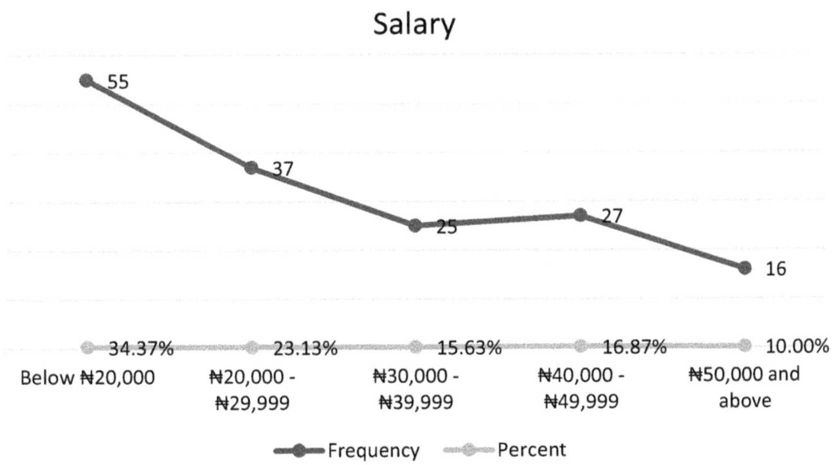

Figure 12.6 Salary of respondents

between ₦40,000 and ₦49,999, while 10% earned ₦50,000 or above per month. With the current exchange rate of one Nigerian naira (₦1) to 0.0028 US$ (US$0.0028), the majority (34.37%) of the respondents earned US$56 per month. This shows that the majority of the respondents were LIEs. With the present high cost of building materials, labour, land and so forth, and the consequences on the cost of housing and rent, LIEs would be deprived of the economic or financial competence to afford liveable and adequate housing.

Figures 12.7 and 12.8 display, respectively, housing types and the percentage of income spent on housing among the respondents. From Figure

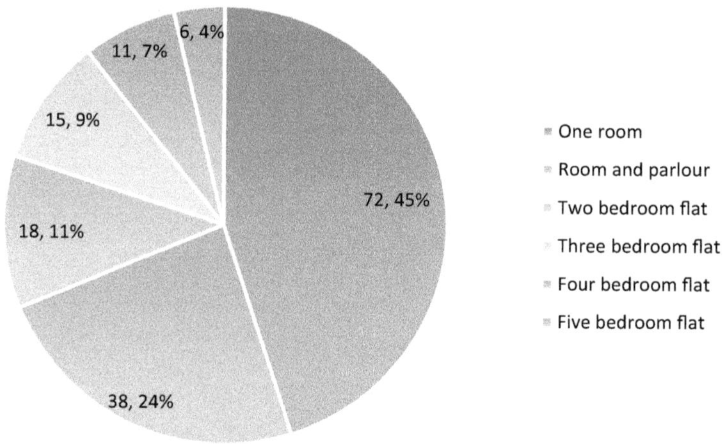

Figure 12.7 Housing types

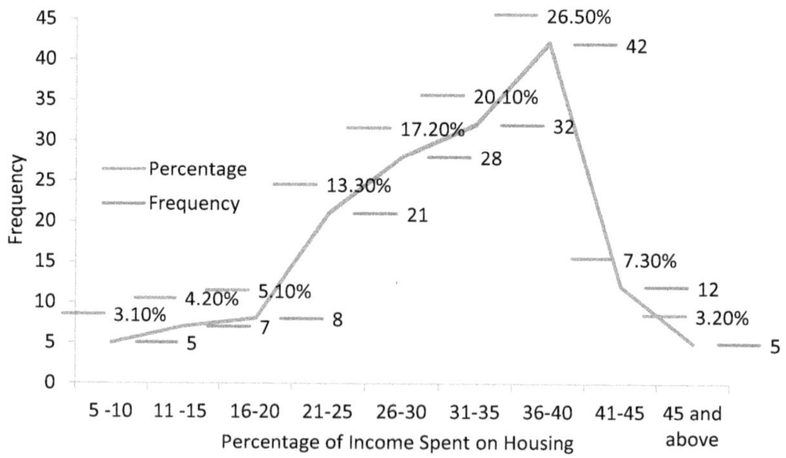

Figure 12.8 Percentage of income spent on housing

12.7, it can be gleaned that the majority of the respondents (45%) lived in one-room apartments. This is followed by those who lived in a room and parlour apartment, which constituted 24% of the respondents. The type of housing that the respondents lived in clearly reflects the consequences of pervasive low income. The majority could not afford adequate accommodation and were instead crowded in one room with their families. Many of the respondents were married (up to 62%; see Figure 12.3), and Nigeria has a high fertility rate with the desired number of children being seven for men and six for women (NPC & ICF International, 2014). This means that the majority of the respondents were likely to have relatively large families. A challenge for large households living in one- or two-room apartments is inadequate sanitation facilities. Many of the buildings are crowded, and the toilet facility is situated separately from the building.

In many cases, several occupants make use of one toilet that is located some distance away from the main building, where occupants must queue and take turns to wash and use the facilities. A lack of adequate space increases the possibility of children abandoning their home for the streets (Heinonen, 2000; Oyefara *et al.*, 2019). It can also give rise to other problems such as a lack of privacy, stress-induced emotional disorders and poor mental well-being (Adugna, 2006; Blum, 2009; Jain, 2007).

Figure 12.8 indicates the percentage of income that the respondents spent on accommodation. The horizontal line displays the percentages. A total of 57.1% of the respondents spent 31% or more of their monthly income on housing. This comprised those who spent 31–35% (20.10%), 36–40% (26.50%), 41–45% (7.30%) and 45% or above (3.20%). The benchmark for affordable housing is expenditure of 30% of household income (McNamara, 2008; UN, 2002). Thus, rent that exceeds this benchmark is considered to be a cost burden, and households may have difficulty obtaining other necessities of daily life such as food, clothing, transportation, medical care and education of children, among others.

5 Discussion

The primary concern of this study was the access to affordable housing and sustainable development among LIEs in Nigeria, with a focus on the Bariga community. Sustainable development entails the ability of individuals or households to meet their present needs and still retain the capacity to meet future needs. However, the capacity to meet one's future needs depends on the sufficiency of one's earnings to cover current basic needs and still have enough to afford future needs and to invest in education, skills acquisition and capital accumulation (UN, 2002).

An important determinant of LIEs' ability to meet other needs is the housing cost burden which is defined as the proportion of monthly income households or individuals pay for their housing costs (McNamara, 2008). Studies have demonstrated that access to housing is central to the quality of people's well-being and life satisfaction, and constitutes a significant predictor of quality of life, while its domain or physical surroundings are said to play a vital role in people's social capital (Habibi & Zebardast, 2018). A high housing cost burden among urban LIEs will affect their capacity to afford other basic human needs and guarantee their future socioeconomic well-being.

In this study, it was shown that the majority of the respondents spent more than 30% of their monthly income on housing; this indicates a high housing cost burden. In general, an appropriate housing cost burden should not exceed 30%. For residents in this situation, sustainable development, which should encompass the ability to earn a living and meet present needs, as well as to make savings for future investments in education, skill acquisition and capital accumulation, would be challenging. A high housing cost burden, combined with a low income, would compromise their capacity to secure the future, and when the burden is severe, it puts people at risk of homelessness (McNamara, 2008). Consequently, the attainment of Goal 1 (i.e. the eradication of poverty in all its forms and building the resilience of the poor) and Goal 11 (sustainable cities and communities) of the SDGs will be challenging in Nigeria.

A similar finding about the high housing cost burden among LIEs had been observed in American society (McNamara, 2008). According to the study, about 79% of LIEs spent more than 30% of income on housing. However, the existence of better PPP and a relatively high minimum wage in American society (US$7.25 per hour, compared with US$0.025 per hour in Nigeria) means that this category of urban dwellers in American society would still have sufficient income to cover some basic needs after paying housing costs. In addition, American society has a system of social security, services for homeless people and shelters to cushion the effects of the chronic housing cost burden (McNamara, 2008). In Nigeria, these facilities are lacking, and the persistent housing cost burden often pushes many people with little or no means of subsistence to live on the street or in open markets and garages.

Furthermore, the study indicates that many of the respondents earned less than ₦20,000 and lived in one-room apartments. In this situation, many would not be able to meet their existing needs, especially those with large families. There is the possibility of poor parenting and upbringing of children. Many families would find it challenging to meet the needs of their offspring, which would thereby cause some children to begin to seek ways of fending for themselves at a very young age. This phenomenon has been reported as the reason why some children abandon their homes to live on the street and become involved in drugs, prostitution, gang violence and associated social vices (Morrison-Valfre, 2013; Oyefara *et al.*, 2019; Siegel *et al.*, 2003). Another concern is overcrowding and its implications for environmental degradation. In other words, beyond the emotional and psychological stress that is generated when too many people are crammed into a limited space, experts have noted that a large concentration of the human population in an area beyond its carrying capacity impacts on the environment negatively; such a crowded environment can be fertile ground for the outbreak and spread of diseases (Fahmi, 2014; Olley, 2006; UN, 2002).

In general, the majority of the people in the study area had inadequate accommodation, and the majority paid above the affordability benchmark. This implies that many of the people in the study area would not have adequate means to procure other necessities such as food, clothing, education, transportation and medical services, among others. Unless efforts are put in place to increase the existing minimum wage in the country and tune the approach to housing provisions in line with the needs of LIEs, attaining the goal of poverty eradication and sustainable communities will remain a mirage among this population.

6 Conclusion and Recommendations

The objective of this chapter was to assess the ability of LIEs to afford liveable housing in Nigeria, and the implications for their present and future well-being. The study was prompted by the concern that the laudable declarations of the SDGs, to eradicate poverty in all its forms and build sustainable cities and communities, will be challenging to attain among LIEs in urban centres because the pattern of housing provision and gentrification by municipal authorities tends to exclude LIEs from the reach of adequate housing. This chapter concludes that the majority of LIEs surveyed were under a high housing cost burden, meaning that many would not be able to adequately provide other basic necessities and save for investments for the future. It is equally clear from the study that many families with a sizeable number of children lived in one-room apartments and crowded environments. This has the potential to affect the psychological and emotional balance of these individuals and give rise to cultural and environmental degradation. The following recommendations are proposed as a way forward:

- Emphasis should be placed on housing units for low- and medium-income earners, made with earth blocks, burnt bricks, compressed earth bricks and standard technology. This will reduce the cost of housing, making it more affordable to many people.
- The National Research Institute should be more empowered to carry out significant research into cheap and effective materials for housing.
- Individuals, especially middle- and low-income earners, should be granted access to long-term credit facilities with soft interest rates. This can be achieved through the establishment of construction or development banks.

- The government should design a policy that will reduce the influx of migrants into cities to reduce the pressure on housing provisions.
- The government should provide alternative strategies for housing construction. For instance, the government could acquire land and design the layout with the necessary infrastructure before making it available for sale to individuals.
- Governments at all levels should provide a favourable investment climate, infrastructure and mortgage insurance to low- and middle-income earners.

References

Adugna, G. (2006) *Livelihoods and Survival Strategies among Migrant Children in Addis Ababa*. Trondheim: Norwegian University of Science and Technology (NTNU).

Akanle, O. (2010) Pentecostalism in Lagos metropolis: Pitfalls and possibilities for national development. In N. Ninalowo, F.A. Badru and R. Akinyemi (eds) *An Interdisciplinary Discourse on the Human Condition* (pp. 151–173). Lagos: Sociology Department, Faculty of Social Sciences, University of Lagos.

Badru, F.A. (2004) Urbanization, family and social change in Nigeria: 1950–1999. In M.O.A. Adejugbe (ed.) *Industrialization, Urbanization and Development in Nigeria: 1950–1999* (pp. 31–52). Lagos: Concept Publications.

Beall, J., Guha-Khasnobis, B. and Kanbur, R. (2010) *Urbanization and Development: Multidisciplinary Perspectives*. Oxford: Oxford University Press.

Blum, W.B. (2009) Young people: Not as healthy as they seem. *Lancet* 374, 853–854.

Chidebell, I. (2013) *Homelessness in Nigeria: Investigating Africa's Housing Crisis*. Bloomington, IN: Xlibris Corporation.

Duranton, G. (2009) Are cities engines of growth and prosperity for development? In M.P. Spence, C. Annez and R.M. Buckley (eds) *Urbanization and Growth* (pp. 67–114). Washington, DC: World Bank.

Fahmi, W.S. (2014) *The Adaptation Process of a Resettled Community to the Newly-built Environment: A Study of the Nubian Experience in Egypt*. Boca Raton, FL: Universal.

Habibi, S. and Zebardast, E. (2018) Exploring the physical-environmental domains of quality of life: The experience of midsize cities in Iran. *Urban Research & Practice* 11 (4), 426–440.

Heinonen, P.M.L. (2000) Anthropology of street children in Addis Ababa, Ethiopia. Thesis, Durham University. See http://etheses.dur.ac.uk/1667/1/1667.pdf?EThOS+(BL)= (accessed 29 November 2019).

Jain, R.K. (2007) *Lifestyle for Total Development: A Unique Guide to Develop your Personality*. New Delhi: Sterling Publishers.

McNamara, R.H. (2008) *Homelessness in America*, Vol. 3. Westport, CT and London: Praeger.

Morrison-Valfre, M. (2013) *Foundations of Mental Health Care* (5th edn). St Louis, MO: Elsevier/Mosby.

NPC (2006) *National Population Census*. Abuja: NPC.

NPC and ICF International (2014) *Nigeria Demographic and Health Survey*. Abuja: NPC.

Olley, B.O. (2006) Social and health behaviours in youth of the streets of Ibadan, Nigeria. *Child Abuse and Neglect* 30 (6), 271–282.

Oyefara, J.L., Eborka, K., Adejoh, S.O. and Akeju, D. (2019) Fostering home return and educational pursuit of street children: Exploring the challenges of a special population. In E. Nwokocha and O. Fayehun (eds) *Concise Demography of Nigeria* (pp. 510 –527). Ibadan: BumbleBee.

Oyekanmi, F.D. (2004) Socioeconomic dimensions of gender equality in Nigeria. In S.O. Akinboye (ed.) *Paradox of Gender Equality in Nigerian Politics* (pp. 41–61), Lagos: Concept Publications.

PRB (2017) *World Population Data Sheet*. Washington, DC: Population Reference Bureau.

PRB (2018) *World Population Data Sheet*. Washington, DC: Population Reference Bureau.

PRB (2019) *World Population Data Sheet*. Washington, DC: Population Reference Bureau.

Siegel, J.S., Welsh, B.C. and Senna, J.J. (2003) *Juvenile Delinquency: Theory, Practice and Law*. Belmont, CA: Thomson Wadsworth.

Tettey, C. (2007) Urbanization in the regions of Africa. In B. Thakur, G. Pomeroy, C. Cusack and S.K. Thakur (eds) *City, Society and Planning* (pp. 163–181). New Delhi: Concept Publishing.

UN (2002) *Sustainable Cities Programme 1990–2000: A Decade of United Nations Support for Broad-based Participatory Management in Urban Development*. Nairobi: United Nations Environmental Programme (UNEP). See https://unep.inadequate+space+and+environmental+environmental%20degradation (accessed 23 August 2019).

13 The Condition of Women in Nigerian Families and National Development: The Imperative for Cultural Re-evaluation and Legal Rebalance

Chinyere C.P. Nnorom and Kennedy Eborka

The thrust of this chapter revolves around the structure of patriarchy and its application for the exploitation, deprivation and subjugation of females in family and community life in Nigeria. In patriarchal ideology, girls and women in many families are viewed as inferior and accorded lesser value relative to their male counterparts. These tendencies find expression in the way in which females are restricted from access to societal resources in terms of the denial of inheritance and successorship rights, a strong son preference, reduced access to education, and other forms of cultural practices that violate their dignity such as cruel widowhood practices. This chapter is theoretical and relies mainly on surveys found within the literature and anecdotal evidence. It argues that cultural practices that do not contribute to women's freedom and socioeconomic progress have no basis to exist. Thus, there is a need for cultural re-evaluation and legal rebalance to shield women from discriminatory practices. Finally, this chapter recommends compulsory education for women up to secondary school level, the strengthening of existing legal codes to protect females from all forms of abuse and exploitation, and the introduction of social security for non-working widows and elderly and unemployed women.

1 Introduction

Across societies in Africa, females are subjected to varying degrees of social and economic subjugation through the instrument of cultural institutions. This has an impact on the socioeconomic impoverishment of women and affects the pace towards national development. The literature on family systems and cultures in African societies in general, and in Nigeria in particular, establishes that female children are commonly discriminated against in the allocation of family resources and access to education in a manner that impinges on their socioeconomic advancement. In some cases, they are starved of adequate parental attention and nutritional requirements in favour of male children (Eguavon *et al.*, 2009; Nzatuzola, 2005; Oyekanmi, 2006). The question that arises from this phenomenon is why female children are discriminated against in many families in Nigerian society. History has demonstrated that, given the opportunity, women can constitute a decisive agent for progressive social change and societal transformation. However, the continuous systematic subjugation of females as inferior across the sociocultural landscape in Nigerian society has pushed many to the fringe of mere existence. Thus, an understanding of the underlying factors propelling the persistence of this phenomenon is essential for the amelioration of the condition of women and to advance the course of national development.

2 Background

As a primary social institution, the family is expected to serve as the building block for the socialisation, development and protection of individuals born into it in preparation for successful participation in wider society. This is the historical and universal responsibility of the family, and it is expected to perform this role without preference for, or discrimination against, any gender. On the contrary, many families in Africa, with particular focus on Nigeria, perpetrate practices that tend to incapacitate female children through systematic exclusion from access to societal resources (Eguavon *et al.*, 2009; Nzatuzola, 2005; Oriola, 2000). This exclusion has detrimental implications for their general well-being, sense of worth and ability to fare as favourably as their contemporaries in societies where females are accorded opportunities to flourish in private and public spaces, such as Europe and North America (Santrock, 2006).

The oppression of women assumes a frightening dimension when they are compelled, under cultural auspices, to withstand dehumanising conditions such as cruel widowhood practices. Other forms of deprivation and indignity suffered by women in Nigeria include the lack of inheritance rights, physical and sexual abuse and ubiquitous accusations of witchcraft (Cheit, 2014). These conditions are created and sustained by patriarchal practices (Ejikeme, 2003; Walby, 1990). In a society with strong patriarchal ideology, the presumed superiority of the male, and the 'insignificance' of the female, is seen in the joy that heralds the birth of a male child. In many cases, the longer it takes to have a male child, the more the anxiety mounts in the family. To illustrate the difference in how male children are valued relative to their female counterparts in Nigeria, most men do not take a second wife if they have only male children in their marriage, but many will hasten to marry another woman if a male child is not forthcoming in their first marriage (Otutubikey, 2004).

The consequences of devaluing women and the associated socio-economic deprivation are huge at both the individual and societal levels. Evidence has demonstrated that people's capacity for the very essence of humanity, the actualisation of the individual's aspirations, is, by and large, tied to the extent to which they are allowed access to societal resources across gender lines without any form of social exclusion (Benneth, 2010). This line of thinking is reinforced in the ideas of freedom and equity as encapsulated in the Universal Declaration of Human Rights. This declaration provides the basis to interrogate the age-old systematic deprivation of the female.

This chapter argues that the oppression of females in Nigerian society and, by extension, African societies, is propelled by the institution of patriarchy. Under cultural auspices, a patriarchal infrastructure is deployed to perpetuate the subjugation of females in favour of males. Attempts at providing an interpretation of the origin of patriarchy have not yielded any convincing conclusion, much less provided appealing justification for its deployment as a tool for the dehumanisation of women. This chapter examines the various dimensions of the deprivation of women within the family through the forces of patriarchy. It directs attention to the link between the discrimination against females and human well-being and national development, with Nigeria as a case study. This is essentially a theoretical paper based on a review of the literature and anecdotal evidence.

3 A Brief Historical Reflection on the Origin of Patriarchy and Attempted Remediation

Globally, it is acknowledged that women tend to suffer certain levels of social deprivation and systematic torment under cultural guises across societies (World Bank, 2007), and the more underdeveloped the society is, the greater the torment. The reactions of several decades to the plights of women across the globe, especially in developing countries (where there seems to be near institutionalised discrimination against the girl child), catalysed the establishment of certain international accords with a declared mandate of ameliorating the condition of women. Notable among these accords are the United Nations Commission on the Status of Women (1946) and the Beijing Conference on Women and the Girl Child (1995). This concern also found expression at the national level in the creation of Ministries of Women's Affairs by many developing countries and the enactment of constitutional provisions to guarantee the rights of women (Oriola, 2000). Despite these overtures, evidence suggests that in many Nigerian families, females as daughters, wives and mothers continue to suffer various forms of cultural molestations under the weight of patriarchal ideology (Eguavon *et al.*, 2009; Ejikeme, 2003).

Scholarly attempts at interpreting the social and historical forces that shaped the development of patriarchy are yet to come to a consensus. Some authors opine that patriarchy derived its force from the phallocentric presentation of social history and its pro-male narratives, such as those found in a segment of humanity's genealogical history from earlier civilisations to contemporary times, which appear to depict man as especially privileged (El-Gilany & Shady, 2007; Otutubikey, 2004). Other scholars, such as Judith Benneth, argue that patriarchy might have a biological origin outside the provinces of history and the social sciences (Benneth, 2010). Similarly, early philosophers such as Plato and Aristotle made allusions to women's natural 'inferiority' and 'defectiveness' (Adekeye, 2008). Furthermore, the books of several religions in the world today are replete with male-privileging narratives that confer superior status on men. This tends to serve as fertile ground for the perpetuation of patriarchy in human society.

Be that as it may, what is not clear is whether the act of presenting a historical narrative in a way that tends to privilege men is sufficient to patriarchalise social existence. This point is relevant because of other aspects of social history that accorded the feminine entity a privileged place in the narrative, but did not propagate the establishment of matriarchy. Take, for instance, the worship of the mother goddess, the goddess of the earth and of all things that live and grow, depicted in female figures in several civilisations, such as in Mesopotamia, Syria, Asia Minor and some parts of West Africa and modern India (Kriwaczeck, 2012). This suggests that the establishment of patriarchy did not necessarily follow from male-privileging narratives of social history, since females were also accorded privileges in the narratives of several civilisations. Thus, the controversies surrounding attempts to account for the origin of patriarchy remain unresolved.

At this point, a conceptual clarification of the notion of patriarchy deserves some attention. In the thinking of Rich (1995), patriarchy is a familial, social, ideological, political system in which men, by force, direct pressure or through ritual, tradition, law and language, customs, etiquette, education and division of labour, determine what part women shall or shall not play, and in which the female is everywhere subsumed under the male. For Benneth (2010), patriarchy can be used to refer to the legal power of a husband/father over his wife, children and other dependents, which lends credence to the idea of patriarchalism or paternalism. Furthermore, Walby (1990) outlined two dimensions of patriarchy to include private patriarchy and public patriarchy. According to the author, in private patriarchy, an individual patriarch (the male head of the household) controls the women directly in the private sphere. It is the man in his position as husband or father who is the direct oppressor and beneficiary of the subordination of women. Public patriarchy represents a situation in which women have restricted access to both public and private arenas. In this case, women are not barred from the arenas but are nonetheless subordinated within them. However, Benneth (2010: 54) was quick to observe that 'patriarchy might be everywhere, but it is not everywhere the same'. This suggests that the deployment of patriarchy as an instrument of oppression varies tremendously from society to society. In the following sections, this chapter attempts to provide an x-ray like image of its manifestations in Nigerian families.

4 Patriarchy and the Condition of Women in Nigerian Society: Dimensions and Manifestations

Social science literature on gender, anthropological and sociological studies in Nigeria is unambiguous about the nature of patriarchy in the society (Bammeke, 2007; Otutubikey, 2004; Oyekanmi, 2006). The phenomenon does not conceal its bias in favour of the male. Thus, in most patriarchal societies, the role of the male as the primary authority figure is clear and remains the basis of social organisation. Males hold unquestioned authority over women and property. And whereas a relationship can be drawn between the mode of traditional authority and inheritance rights in a patriarchy, the same cannot be said of matriarchy in many African societies. For instance, in Angola, southern Africa, 'household assets are generally the property of male heads of households, and inheritance usually benefits the male relatives of the deceased' even though the society is traditionally matrilineal in most ethnic groups (Nzatuzola, 2005: 109). Similar practices have been reported among the Ohafia and Afikpo people of Southeastern Nigeria (Nnorom, 2010). In other words, even in societies in Africa where matriarchy is the rule, inheritance rights still favour the male. In Nigerian society and its family system, the forces of patriarchy dictate the manner in which society relates to women, such that even the unborn female is already fated to be encircled by pre-established cultural and economic restrictions (Igwe & Akolokwu, 2015).

4.1 Patriarchy and the condition of women in Nigerian society: The sociocultural dimension

The bias against females in many families across Africa is clearly expressed in male child preference and the consequent exclusion of females from access to societal resources necessary for overall enhancement and creative fulfilment. In many families across Nigeria, the preference for the male child is clearly evident through discrimination against females in the allocation of family resources and familial attention (Ejikeme, 2003; Otutubikey, 2004; Oyekanmi, 2006). An examination of social practices in everyday life among many Nigerian families shows instances of unequal treatment of females, relative to males. In many families, the birth of a male child is often heralded with greater joy than that of a female. The presence of at least one male child is regarded in many families as absolutely necessary, and some couples go to the extent of consulting oracles and religious prayer warriors to ensure the birth of a male child. The bias in favour of males also finds expressions in the onomastics of Igbo and Ibibio names. Among these groups, male children are named to represent strength, hope and fulfilment, and the longer it takes to have a male child, the more anxiety mounts in the home. When the male child finally arrives, he is given, with great relief, a name such as Obisike, Obidike (my heart is strengthened), Ebisike (I now live stronger), Nwokedi (a man is here), Obiajulu (my fear has calmed) and so on (Amadiume, 1999). Until the arrival of a male child, the names of female children will boldly declare the crisis of anticipation in the family. Examples of such names are Ogadinma (it will be better soon), Ndidi (patience), Anaelechi (I hope in God), Anaedi (I continue to persevere). This practice depicts the male as an important figure of great social and economic investment.

Patriarchy also finds expression in the nature of customary laws on succession in many societies in Nigeria. It is customary for daughters to be excluded from successorship and inheritance rights in many cultures. By nature, customary law is basically unwritten, unlike constitutional legislation, and represents the embodiment of the rules, customs and practices of a people that have been given statutory and judicial validations in the course of their existence (Oriola, 2000). Given that it originates from customs, its application tends to vary from culture to culture. Under the customary law of succession in Nigeria, if a man dies intestate (without a will), properties are shared in accordance with the customary law of the society. For instance, if a Muslim man dies intestate, his estate must be shared among his heirs entitled to a share under Islamic law. His male children will each have equal shares and the female children a half share. His widow is also entitled to some portion. In contrast, in many cultures in southern Nigeria, a widow cannot inherit her husband's property, although her children can, depending on the sex of the child and the custom of the society. Among the Igbo, under the Onitsha customary law, a widow cannot inherit her husband's property, and if the man died without a male child, the property would pass to the deceased's relatives. A similar practice is found among the Bini of Edo State, Nigeria (Oriola, 2000).

Unlike the Bini and Igbo customs, where widows are denied successorship rights to their deceased husbands' property because they have only female children, the Yoruba custom allows daughters to inherit the property of their deceased father (Agomo, 2004; Oriola, 2000). In a study among the Yoruba of Ondo and Ekiti States, Nigeria, it was observed that female children are taken into consideration in the sharing of the family's property. Nonetheless, the majority (about 80%) of the widows interviewed in the study revealed that they were denied access to the property left behind by their late husbands. Consequently, there is a dichotomy between what is customarily stipulated and what is actually practised in these societies (Aderinto, 2004; Agomo, 2004). In general, the nature of customary law in most societies in Nigeria reflects a sociocultural system that is under the grip of patriarchy and overwhelmingly anti-female. What this means for women is that they are a rejected species left to fate. In her family of orientation, a woman is denied rights of inheritance; in her family of procreation, she is denied rights of succession if she had no male child. These practices put the woman and her female children in a severe socioeconomic condition.

The maltreatment of women is not limited to the restrictions they suffer in terms of access to social and economic development; there are also various forms of molestation and dehumanisation that take place under cultural guises. In some cultures in Nigeria, when a man dies, his widow becomes an object of suspicion, especially by her late husband's family, for responsibility for his death. This is particularly so if it is termed an untimely death. Even when the death is not conceived of as untimely, the widow is expected to renounce her rights to a dignified and decent existence as a sign of genuine bereavement. Among the Edo of Southsouth Nigeria, for instance – in addition to the instant and compulsory isolation and enforced wailing – the widow is required to undergo other inhumane treatments, such as eating from unwashed plates for a period of three to four months or forced shaving of the head, and is sometimes made to drink the water used to bathe the body of her dead husband (Eguavon *et al.*, 2009; Genyi & George-Genyi, 2013). While these practices persist, it has been observed that many of the women who suffer these treatments the most are those in weak socioeconomic positions (Igwe & Akolokwu, 2015). In general, whereas women are expected to go through deplorable widowhood practices in mourning for their deceased husbands, men are barely required to undergo any form of extended mourning ritual on the death of their wives.

Other conditions suffered by females in many African families, which might owe their roots to the patriarchal social structure, include the objectification of girls and women as items of sexual abuse and domestic violence (Abiad & Masoor, 2010; Bammeke, 2007; Eguavon *et al.*, 2009; Nnorom, 2010). In the northern part of the country, under-age girls are forced into marriage under religious and cultural dictates. One such case, in Kano State, led a child bride, Wasila Umar (née Tasiu), a 14-year-old girl, to poison her 34-year-old husband, Umar Sani, who was already married with two children. According to the report, Wasila Umar was forced by her father to marry Umar Sani against her will. Two weeks after the marriage, the child bride, in a desperate bid to free herself from the forced marriage, added rodenticide (rat poison) to her husband's food to eliminate him. Unfortunately, 14 other persons joined her husband in eating the food, and three of them, including Wasila's husband, died immediately after the meal (BBC, 2014). On domestic violence, a report from the Nigeria Demographic and Health Survey (NPC & ICF International, 2013) shows that 28% of all women had experienced physical violence since the age of 15, and many of these acts of violence resulted in injuries to the victims.

There are also concerns about the subtle violence against females expressed through female genital mutilation (FGM). Female genital mutilation or cutting (FGM/C) is the controversial practice of partial or total excision of young girls' genitals for nontherapeutic reasons, in than the belief that it helps to promote chastity and prevent sexual promiscuity (WHC, 2012). It is now internationally accepted that FGM/C is a harmful cultural practice and an act of violence against the physical and emotional integrity of women which directly infringes upon women's (and children's) rights to physical, psychological and social health wherever and however it occurs. The World Health Council has attempted to highlight some of the health and social repercussions of FGM/C. Short-term issues include death, the threat of death from haemorrhaging, and shock from the pain and level of violence associated with carrying out the procedure. The long-term effects include urinary tract infections and increased pain during menstruation, as well as the development of scar tissue and difficult labour and delivery (WHC, 2012). Despite these findings, the procedure is still performed in many families in Nigeria and other parts of Africa, the Middle East and Asia (NPC & ICF International, 2013; UNICEF, 2005; WHC, 2012).

The narrative that tends to link the phenomenon of male child preference to patriarchy has been questioned by some scholars who argue that male child preference is not necessarily an offshoot of patriarchy, but a manifestation of a society's sense of valuation and reproductive preferences (El-Gilany & Shady, 2007). They observe that although the phenomenon is most widespread in poor countries where the status of women is low, one should not make any hasty generalisation that patriarchy inevitably engenders male child preference. Making reference to the 2006 World Fertility Survey (WFS) report, they argue that Indonesia and a number of South American societies are patriarchal, yet they have equal preference for both male and female children, lending credence to the assertion referred to earlier that 'patriarchy might be everywhere, but it is not the same everywhere' (Benneth, 2010: 54).

4.2 Patriarchy and the condition of women in Nigerian society: The socioeconomic dimension

During the period around the 1980s, a number of women's rights groups and feminist movements in the West appeared on the scene in what seemed like counterhegemony to patriarchal forces. Until then, more than 90% of women in the world were consigned to the background and rendered voiceless in almost every sphere of social life as a result of rigid patriarchal social systems (Moghadam, 2007). In recent decades, while things appear to be undergoing some changes in response to the pressure and activities of women's rights groups in the West, in much of Africa, the Middle East and South Asia, which Moghadam describes as a 'patriarchal belt', patriarchy has been resisting the forces of change with dire socio-economic implications for many women.

In many Nigerian cultures and families, for instance, patrilineal rule dictates that a widow without a male child automatically forfeits every moral, cultural and legal right to her deceased husband's property if he dies intestate, irrespective of how many female children she had with her late husband. Consequently, she and her children are doomed to a severe economic situation. This practice has been noted as being responsible for the pervasive cruel conditions that many women in African families experience and as a reason why women predominate among those suffering severe economic and social miseries. Many divorced women, widows and female-headed households suffer extreme poverty. Patriarchy has propagated a system in which women in the family are systematically deprived, thereby making them more vulnerable to negative socioeconomic conditions (Igwe & Akolokwu, 2015).

The weight of patriarchy upon women in many Nigerian families is equally evidenced by the asymmetry in access to societal resources between males and females. Data from the Population Reference Bureau (PRB) on World's Women and Girls for Nigeria indicate that 39% of females aged 15 years and above were economically active, compared to 73% of males of the same age (PRB, 2011). In Sub-Saharan Africa, the participation of women in tertiary education is less than half that of men, and less than 1% of women have tertiary education (El-Gilany & Shady, 2007). This asymmetry in the allocation of societal resources is sometimes justified by patriarchal apologists as indicative of the inherent superiority of male over female. This claim, however, is questionable as historical evidence demonstrates clearly that, given the opportunity, women can be a decisive force for societal transformation.

4.3 Narratives and counter-narratives on male superiority and women's inferiority

Among several suggestions as to why women are consigned to low status and the associated subjugation in family and community life is the depiction of men as 'superior' and women as 'inferior'. This depiction arose from slanted narratives of social history, which concentrate more attention on one aspect. This pattern of narrative was later to assume prominence in the cultural fabric of many societies in Africa (Blackwell *et al.*, 2008; El-Gilany & Shady, 2007). The designation of men and women as 'superior' and 'inferior', respectively, derived from the social interpretation of men as 'strong' and women as 'weak'. 'Strong' or 'weak' in this sense do not necessarily connote the extent of physical agility, but are to be understood from mental and moral perspectives and tend to blame the female for all human woes, even in circumstances where the male is equally guilty (Blackwell *et al.*, 2008; Klosko, 2012). This narrative is often encountered in the presentation of preserved folklore and legends, and in major religious books. For instance, in the book of Timothy in the Bible there are declarations that appear to put women under men's subjugation based on the idea that it was the woman who was deceived by the serpent to eat the proverbial forbidden fruit and thereby transgress against biblical injunctions. That account, it could be said, attempts to exonerate the man from any wrong, with the claim that it was not Adam that was deceived (1 Timothy 2:13–14). Then, in 1 Peter 3:7, women are expressly pronounced as 'weaker vessels'. The question is, if the configuration of woman as inferior to man is validated on the basis of her encounter with the serpent and her munching of the forbidden fruit, does it not stand to reason, therefore, that Adam was equally inferior and by no means superior to his wife by his subsequent involvement in the consumption of said fruit? Otherwise, why was he not able to resist eating the fruit if he were of any superior capability?

Putting religious literature to one side, a glimpse through ancient Japanese and Greek literature shows that these societies regarded women as inferior to men (Klosko, 2012; Takemaru, 2010). The writings of neo-Confucian scholars portray women as inferior and maintain that 'disobedience, discontentment, slander, stupidity and jealousy are the five worst disorders that torment and make women inferior to men' (Takemaru, 2010: 13). Similarly, in his history of political theory, Klosko (2012) shows that women were regarded as inferior to men and treated accordingly in ancient Greek society. The author maintains that, although women were citizens, they had limited legal and political rights; they were unable to participate in political affairs and were barred from owning property. Their main function, it was observed, was to produce children as heirs for their husbands. This represented the popular view about women in ancient Greek society and dominated much of their social and historical narratives until the emergence of Plato and the counter-narratives in subsequent eras.

The counter-narrative on this subject draws attention to the fact that the grounds that were used to build the hypothesis of males' superiority are not specific to men. Its goal is to deconstruct the obscurantist's perspective on the status of women, and this was boosted by the thinking of major earlier philosophers such as Plato.

Plato's thinking departed from the earlier view on women and he argued that women should have the same opportunities as men to be educated in philosophy and to rule the Just City. In Plato's view, women were sufficiently qualified and capable of ruling the city-state. He was said to have declared that 'the qualities required in a philosopher are not affected by sex, just as the difference between being bald or having long hair does not affect one's ability to serve as a carpenter' (Klosko, 2012: 94). According to Plato, none of the differences between men and women, such as the fact that some men are generally stronger physically, bears directly on women's abilities, because there are also some women who are stronger than some men. This sums up Plato's ideas on community and family. Subsequent work on the counter-narratives had continued to grow, with Blackwell *et al.* (2008: 184) arguing that 'women are experts in things men just don't understand, and that women are both different from and superior to men'. Given the opportunity, women have proven themselves to be as capable as men in almost all areas of human endeavour in recent history.

Furthermore, there are historical examples that challenge the assumption that men are, in general, physically superior to women. The example of the Kingdom of Dahomey (the modern-day Republic of Benin), with its history of a female military regiment and warriors, illustrates the weakness of the assumption of women's inferiority based on physical capability. These legendary female warriors were consistently judged to be superior to male soldiers in terms of effectiveness and bravery (Edgerton, 2000). An addition to the ongoing counter-narratives is Anyiam Osigwe's submission on the pluralistic composition of Divinity and the multiplicity of the Godhead. In Osigwe's submissions on the treatise 'Women as the Salt of the Earth', he asserts that Divinity is a component of the feminine and masculine principalities, and that the wholesomeness of the God principle is in these two principalities mutually coexisting as the total of the Godhead (Ushigiale, 2004).

In summary of this section, it is possible to counter existing narratives that portray females as inherently inferior to males, and stress that the supposed inferior status consigned to women derives from the pattern of presentation of social history which established patriarchal institutions and ideology to the disadvantage of females.

5 Women's Subjugation and Poverty in a Globalising World

The UN made the eradication of poverty in all its forms its first goal in the Sustainable Development Goals (SDGs) agenda. This not only highlights the undesirability of poverty but also indicates the UN's acknowledgement of the significant role poverty plays in the individual's ability to achieve broader social and economic goals and its negative impact on the larger society. However, this laudable goal will be difficult to achieve without a gradual turning away from institutions that perpetuate poverty in societies such as Nigeria. The problem of poverty cannot be addressed by mere cash or material handouts. The issues of management and the mechanisms of access to resources are vital to poverty reduction. When a section of a population is oppressed through cultural imperatives and orientation, they tend to accept their condition as a matter of fate in what could be described as a form of 'learned helplessness'.

Through the instrument of patriarchy, females are abused, demonised and denied access to societal resources. To that extent, even when a society is endowed with material resources, patriarchy precludes women's free and sufficient participation in the development processes of the society by dictating who gets what at the societal and tribal levels. This explains the predominance of women among the world's poorest people (World Bank, 2007). Reports indicate that 80% of those living in extreme poverty (defined as those living on less than US$1.90 per day) live in South Asia and Sub-Saharan Africa, and women are more likely to be poor than men because they lack opportunities (UN, 2016). In Nigerian society, across the socioeconomic and political landscape and also the public space, the visibility of women is dim and disproportionate to their number and significance in societal development. Gender disparities are evident in formal sector employment, which is heavily dominated by men. Women are predominantly engaged in agriculture and are far less likely than men to be engaged in professional, technical and managerial fields, just as they lag behind in educational attainment, literacy and exposure to mass media (NPC & ICF International, 2013). The agricultural sector, by all indications, has remained largely crude over the decades and women participate as unpaid family workers, self-employed or wage earners. In addition, land is vital as a result of the importance of agriculture in Nigeria's economy, but women only own or have access to a fraction of titled land as a result of cultural impediments.

In Nigeria's political arena, the proportion of seats held by women in national parliaments since the return to democracy has been abysmal. The highest record was 7% in 2010, but this dropped to 5.6% in 2017. In 2018, out of 109 Senate seats, women occupied only seven, and in the lower house, the House of Representatives, they held only 22 out of 360 seats. Yet it has been acknowledged globally that sufficient participation of women in socioeconomic and political processes is linked to national development in terms of improved policy changes, economic growth, enhanced peacebuilding and a more egalitarian society (World Bank, 2007).

Commentaries on the reasons for these disparities vary. While some argue that the disparity is the result of the violence and troubles that characterise Nigeria's political processes, others contend that women do not have the capacity to participate in public life and engagements. The aim of this chapter is not to contend with any position. However, a keen observation of the two positions would reveal the influence of patriarchal institutions and ideology. On the whole, when a large segment of a society's population is subjugated and deprived of access to societal resources, their opportunity for self-development is confiscated. Their ability to participate in today's globalising world for their benefit is affected. When we consider this situation with regard to women, the consequences on society are huge because of their childrearing responsibilities and the training that women provide to younger members of society.

5.1 Towards cultural re-evaluation

Re-evaluation involves the re-examination of a phenomenon or event in order to judge its value, benefits and practicality. For our purposes, cultural re-evaluation is an attempt to interrogate the culturally transmitted blueprint of social relations that typifies the female as inferior and insignificant (Brock, 2000). As a concept, culture refers to the way people act, the body of tradition, the rituals and beliefs that people have learned as members of society, as well as the continually changing pattern of learned behaviour that is shared and transmitted by members of a social group (Blackwell *et al.*, 2008). Within the context of social dynamics, this means that the perception that women are inferior is not a natural state of things, but a culturally constructed idea malleable by the forces of social change. In light of this, it would be safe to argue that the ascription of negativity to girls can equally be deconstructed.

5.2 Towards the reinforcement of legislative provisions

Women's rights to equal economic opportunities are upheld by the International Convention on the Elimination of All Forms of Discrimination Against Women (CEDAW) which came into being in 1979 and was ratified by Nigeria on 13 June 1985. Among other goals, Article 13 of the Convention requires member states to enact measures to eliminate discrimination against women in all spheres of social life. In more specific terms, Articles 11, 14 and 15 state that women have equal rights with men in areas of employment, education, access to credit facilities, choice of profession and access to land (UN, 2012).

At the national level, many African countries have attempted to implement laws to guarantee women's rights as stipulated in Section 42 (1a–b) of the 1999 Constitution of the Federal Republic of Nigeria. It states that no citizen of Nigeria should be discriminated against or subjected to disabilities or denied privileges on the basis of sex, religion or place of origin. Nonetheless, ingrained traditional ideas about the inferior status of women are still prevalent and continue to hamper the implementation of these provisions. Consequently, girls and women suffer severe disadvantages compared to men in access to family and community opportunities, which have an effect on their lifelong prospects. The existing legal code in the 1999 Constitution of Nigeria is toothless and insufficient to guarantee the rights of women that are embodied in the Universal Declaration of Human Rights (UDHR). To this end, there is a need for legal rebalance to shield women in Nigerian family and community life against fates propelled by discriminatory cultural practices. This will enable them to pursue their aspirations and happiness without hindrance.

6 Conclusion and Recommendations

Patriarchy describes a form of social organisation in which authority at the family and community levels is concentrated on males. This has been observed to be responsible for the systematic exclusion of women in many African societies from access to societal resources and their resultant widespread poverty. The view that males are superior to females can be contested on account of history and recent developments around the world. The weakness of this view is exposed when the achievements of women in various fields in recent times are considered. Given the opportunity, the female population can constitute a formidable force for society's transformation. The following are suggestions for a way forward towards the socioeconomic and cultural emancipation of females and Nigerian society as a whole, based on the content of this chapter:

- There should be explicit constitutional legislation at national, state and local levels to protect girls and women, specifically, from all forms of exploitation, discrimination and cultural molestation. Such legislation should be designed to override customs that fail the repugnancy, equity and good conscience tests in modern jurisprudence across societies in Nigeria in order to shield girls from discriminatory practices imposed by customs.
- Enlightenment programmes aimed at educating females on their constitutional and civil rights (and how to exercise these rights) should be initiated and extended to the rural areas where primitive cultural dictates are still unquestioned and strong.
- Education for females should be made compulsory to the end of senior secondary school. Many should be encouraged through various incentives to further their education to tertiary level. This will help to demystify the myth about the inferiority of females and shield them against inhumane cultural practices. To this end, there should be a provision of special funds for indigent females who wish to go to school but lack the financial resources. Similarly, social security should be introduced for non-working widows and elderly and unemployed women.
- Finally, there should be value reorientation at the family level. This can be supported by civil society organisations and advocacy groups, with the collaboration of the National Orientation Agency of Nigeria.

References

Abiad, N. and Masoor, F.Z. (2010) *Criminal Law and the Rights of the Child in Muslim States: A Comparative and Analytical Perspective*. London: British Institute of International and Comparative Law.

Adekeye, O.A. (2008) Prevalence and patterns of gender violence: Major variables in the exposure to HIV/AIDS among women in Nigeria. *Gender and Behaviour* 6 (2), 127–140.

Aderinto, A.A. (2004) Widowhood practices among the Yoruba of Ondo and Ekiti States in Nigeria. *Journal of the Nigerian Institute of Social and Economic Research* 20 (1), 66–95.

Agomo, C. (2004) The legal framework for gender equality in Nigeria. In C. Akinboye (ed.) *Paradox of Gender Equality in Nigerian Politics* (pp. 91–100). Lagos: Concept Publications.

Amadiume, I. (1999) *Male Daughters, Female Husbands: Gender and Sex in an African Society*. London: Zed Books.

Bammeke, F. (2007) Gender, household headship and the cultural undertone: Illustration from Nigeria. *Unilag Sociological Review* 8, 21–56.

BBC (2014) *Nigeria child bride poisons older husband*. Online: https://www.bbc.com/news/world-africa-26978872.

Benneth, J.M. (2010) *History Matters: Patriarchy and the Challenge of Feminism*. Philadelphia, PA: University of Pennsylvania Press.

Blackwell, J.C., Smith, M.E.G. and Sorenson, J.S. (2008) *Culture of Prejudice: Arguments in Critical Social Science*. Toronto: University of Toronto Press.

Brock, D.W. (2000) Broadening the bioethics agenda. *Kennedy Institute of Ethics Journal* 10 (1), 21–38.

Cheit, R.E. (2014) *The Witch-hunt Narrative: Politics, Psychology, and the Sexual Abuse of Children*. Oxford: Oxford University Press.

Edgerton, R.B. (2000) *Warrior Women: The Amazons of Dahomey and the Nature of War*. Boulder, CO: Westview Press.

Eguavon, A.N., Odiagbe, S.O. and Obetoh, G.I. (2009) The status of women, sex preference, decision-making and fertility control in Ekpoma Community of Nigeria. *Journal of Sociology* 15 (1), 42–49.

Ejikeme, G.G. (2003) *Socioeconomic and Cultural Conditions of the Girl Child*. Jos: Decca Publications.

El-Gilany, A.H and Shady, E. (2007) Determinants and causes of son preference among women delivering in Mansoura, Egypt. *Eastern Mediterranean Health Journal* 13 (1), 15–39.

Federal Republic of Nigeria (1999) *1999 Constitution of the Federal Republic of Nigeria*.

Genyi, G.A. and George-Genyi, M. (2013) Widowhood and Nigerian womanhood: Another context of gender poverty in Nigeria. *Research on Humanities and Social Sciences* 3 (7), 68–73.

Igwe, O.W. and Akolokwu, G. (2015) Patriarchy and its violations of human rights of women in Nigeria. *International Law and Policy Research Journal* 1 (1), 001–008.

Klosko, G. (2012) *History of Political Theory: An Introduction, Vol. 1. Ancient and Medieval* (2nd edn). Oxford: Oxford University Press.

Kriwaczeck, P. (2012) *Babylon: Mesopotamia and the Birth of Civilization*. New York: St Martin's Griffin.

Moghadam, V. (2007) Women's empowerment: An introduction and overview. In V. Moghadam (ed.) *From Patriarchy to Empowerment: Women's Participation, Movements and Rights in the Middle East* (pp. 1–16). Syracuse, NY: Syracuse University Press.

Nnorom, C.C.P. (2010) Sexual abuse of children and the Child Rights Act in Nigeria: Is implementation feasible? In F.A.D. Oyekanmi, F.A. Badru and F. Bammeke (eds) *Globalization and Social Pathologies: Rethinking the State of Social Welfare in Nigeria* (pp. 222–236). Lagos: Department of Sociology, University of Lagos.

NPC & ICF International (2013) *Nigeria Demographic and Health Survey 2013*. Abuja: National Population Commission.

Nzatuzola, J.B. (2005) Gender and family life in Angola: Some aspects of the post-war conflict concerning displaced persons. *African Sociological Review* 9 (2), 106–133.

Oriola, A. (2000) Widows' right of inheritance under Nigerian Customary Law: A discourse. *Unilag Sociological Review* 1, 142–158.

Otutubikey, I.C. (2004) *Patriarchal Ideology and Discourses of Sexuality in Nigeria*. Uyo: Africa Regional Sexuality Resource Centre, Department of Sociology and Anthropology, University of Uyo.

Oyekanmi, F.A.D. (2006) Language, culture and national development. In T.A. Ezeigbo and K.K. Aribisola (eds) *Literature, Language and Nature Consciousness* (pp. 81–103). Lagos: University of Lagos Press.

PRB (2011) *World Population Data Sheet*. Washington DC: Population Reference Bureau.

Rich, A. (1995) *Of Women Born: Motherhood as Experience and Institution*. New York: W.W. Norton.

Santrock, J.W. (2006) *Life Span Development*. New York: McGraw-Hill.

Takemaru, N. (2010) *Women in the Language and Society of Japan: The Linguistic Roots of Bias*. New York: McFarland.

UN (2007) *World Fertility Report 2007*. United Nations Department of Economic & Social Affairs. Online: https://www.un.org/en/development/desa/population/publications/pdf/fertility/worldFertilityReport2007.pdf

UN (2012) *Treaty Collection*. See https://treaties.un.org/Pages/Home.aspx?clang=_en (accessed 26 February 2015).

UN (2016) *Sustainable Development Goals Report 2016*. See https://unstats.un.org/sdgs/report/2016/.

UNICEF (2005) *Female Genital Mutilation/Cutting: A Statistical Explanation*. New York: UNICEF.

Ushigiale, J. (2004) After Peres, who next? *Thisday Newspaper*, 16 October, 10 (3463), 15.

Walby, S. (1990) *Theorizing Patriarchy*. Oxford: Blackwell.

WHC (2012) *Female Genital Mutilation/Cutting: A Literature Review*. London: Women's Health Council.

World Bank (2007) *Gender and Economic Growth in Tanzania: Creating Opportunities for Women*. Washington, DC: World Bank.

14 Cuban Tourist Enclaves: A Critical Approach

Renata Rettinger

Cuba is one of the fastest developing tourist regions in the Caribbean. This dynamic growth in tourist traffic is caused by increased investment in the tourism infrastructure. Since the 1990s a trend in the tourist infrastructure could be observed, where building all-inclusive hotels in the coastal region has become increasingly popular. In Cuba's case these were usually closed areas – tourist enclaves, where only foreign tourists and Cuban staff could reside. Thus the tourist regions were located in isolated places such the islands of Cayo Coco and Cayo Largo del Sur. The inception of the Cuban tourist enclaves was caused by two factors: isolating tourists in order to better control them, and isolating Cubans from Western living standards and lifestyle. The goal of this chapter is to critically analyze the development of new Cuban enclaves as well as those already existing. The elaboration considers both quantitative and qualitative problems, which are analyzed spatially in the polo turistico system. The issue will be presented against the backdrop of other tourist regions across the globe. In addition, this chapter aims to identify the present and potential environmental, societal and economical conflicts brewing in the tourist regions, in the context of sustainable development. Field studies conducted in 2010 and 2018 are the basis of this chapter.

1 Introduction

Cuba is one of the fastest growing tourist regions in the Caribbean, where the rapid growth of its tourism sector is driven by multiple factors including the quality of its tourist infrastructure. The trend since the early 1990s has been to build large hotel complexes offering all-inclusive services along Cuba's coastline (Apollo & Rettinger, 2019; Kowalczyk & Derek, 2010; Padilla & McElroy, 2007; Rettinger & Wójtowicz, 2014). The all-inclusive model of tourism development has been predominant in the Caribbean for years (Issa & Jayawardena, 2003; Rettinger & Apollo, 2017; Rettinger & Wójtowicz, 2014) and cruising is a major part of this trend – and both are related to the enclave tourism concept. It has been shown that this type of tourism is a major reason for choosing Caribbean travel sites.

The leading features of this variety of tourism include the beauty of the natural environment, a 'tropical' climate, non-material cultural heritage, the presence of all-inclusive and highly ranked hotels, excellent transportation options, hotel security and beach security (Lee & King, 2006).

The above statements suggest that the Caribbean is a region where tourists value exclusive recreational options. This is based on average daily expenditure per tourist in the Caribbean region (Theuns, 2011). According to data published by the Caribbean Tourism Organization (2016), the number of foreign tourists per country are: Dominican Republic (6.7 million), Cuba (4.0 million), Jamaica (3.8 million), Puerto Rico (3.7 million) and Bahamas (1.4 million). Caribbean economies are some of the most dependent on tourism (Issa & Jayawardena, 2003). As in the case of the entire region, Cuba relies strongly on tourism, but political reasons preclude cruising. The basis of the Cuban tourism economy consists of recreational tourism along the Cuban coast at highly ranked hotels.

Tourist enclaves are self-sufficient tourist centers and are completely isolated from the local community and function independently of it. Most enclaves are highly dependent on the influx of foreign capital and the current situation in markets generating tourist traffic. Enclave isolation leads to low multiplier effects. At the same time, enclaves are highly self-sufficient in a physical, social and economic sense, and are not effectively linked to the local market. An enclave as a spatially concentrated form of tourism development may consist of one hotel (Derek, 2007), a complex of hotels, or an entire area as in the case of Cuba. The model of enclave tourism development in the Caribbean is generally perceived as a very safe form of tourism, which further drives its development. The security of tourists is a priority issue, as any type of incident involving foreign tourists may negatively affect regional economic development due to the high degree of dependence on the positive image of an area.

Institutional assistance is a major factor driving the emergence of tourist enclaves, as a proper formulation of regulations aids in the development of uninhabited and environmentally valuable areas and includes enclaves in local development strategies. Spatial concentration of tourist infrastructure is one feature of tourism in Cuba; concentration of capital in the tourist sector is another. This is manifested in the fact that Cuba's government owns a majority share in the tourism sector via its largest company – Cubacan – which controls most of the tourism market in the country.

2 Materials and Methods

The principal purpose of the study was to characterize the main trends in the tourist enclave development area in Cuba in the broader context of spatial diversification of tourist traffic. The chapter provides a number of case studies and discusses current actions including those focused on increasing the number of tourists and the quality of the tourist experience.

It further attempts to evaluate key factors, actions and consequences related to local geographic environments in the context of long-term development. The study also provides a review of the offerings of the largest international tour operators in order to determine the principal profile of each company's offering. Given that all the tourist service packages examined in this study consisted of all-inclusive travel deals, the chapter provides a comparison of price points and package durations with respect to each studied tourist region.

3 Results

One very important spatial aspect of contemporary tourism is the establishment of key tourist regions characterized by significant environmental value as well as very good tourism infrastructure and transportation accessibility. This approach leads to a strong concentration of tourist traffic, which may yield both positive and negative effects. The concentration effect is used in the regional development strategies employed across Cuba, which is home to many major tourist regions (*polo turistico*): Havana, Varadero, Jardines del Rey, northern coast of Camagüey, northern coast of Holguin, Santiago de Cuba, southern and central coastal region and Archipelag Los Canarreos. The regional development policy of Cuba's government is based strongly on the primacy of the Havana and Varadero regions. Other key tourist regions in Cuba are Cayo Coco and Cayo Santa Maria. The government of Cuba has made an effort to move some tourist traffic to regions other than the leading region of Varadero. The shift has been to the tourist enclaves of Cayo Coco and Cayo Santa Maria.

The research literature sometimes uses the term 'aggressive' to describe the growth of tourism in Cuba. This is evident in the number of hotel beds, especially at four-star and five-star hotels. According to Cuba's Ministry of Tourism, it is expected that the number of hotel beds will increase by about 3000 per year in the next few years. The pattern of concentration in four key regions will continue: Havana, Varadero, Sabana–Camagüey and north Holguin. In early 2000, the government's plan was to increase the number of hotel rooms: Varadero – from 15,200 to 27,100; Havana – from 10,400 to 36,600; Jardines del Rey and Camagüey – from 5200 to 52,300; Holguin – from 4500 to 15,300. Additional features to be added to the tourist offering are three or four golf courses and three marinas and seaports. According to the Ministry of Tourism, there should be one golf course per roughly 2000 hotel rooms.

The main reason for the establishment of tourist enclaves in developing countries is to substantially isolate foreign tourists from the local population, which is supposed to guarantee security and convenience for the visitors. In the case of Cuba, this isolation is also supposed to isolate Cubans from the lifestyles and living standards of Western tourists – along with the more 'traditional' purpose of isolation of tourists from the everyday lives of local residents (Cerviño & Cubillo, 2005; Matthews, 1977; Padilla & McElroy, 2007). It was decided that tourist enclaves should be established first and foremost at isolated sites – geographically isolated sites such as islands. Island archipelagos are found across the northeastern (Sabana–Camagüey) and southwestern (Archipelag de Los Canarreos) parts of Cuba. The decision to isolate tourists from locals was made easier not only by natural geographic factors but also by the lack of permanent settlements in these regions. However, it was the environmental value of these regions that made them especially attractive to tourists: sandy beaches, warm water and a pleasant climate. One important tourist enclave consists of the Sabana–Camagüey Archipelago, which includes 2500 islands stretching across a distance of 450 km from Punta Maternillos on Cayo Sabinal Island in Camaguey province to Punta Hicacos on the Varadero peninsula located in Matanzas province. One part of this archipelago is known as Jardines del Rey, which is a group of islands about 200 km in length. The most developed islands in this archipelago are Cayo Coco, Cayo Guillermo and Cayo Santa Maria. Cayo Coco Island has an international airport, and is linked to the Cuban mainland via a 27 km causeway. The island is home to ten hotels with a total of 3483 rooms. It is expected that the number of hotels will increase by 50, to a total of 20,000 rooms. Facilities in the region also include a marina and a golf course, with two more under construction. Cayo Coco is linked with Cayo Guillermo – an island with an area of 13 km^2 whose facilities consist of four hotels with a total of 1093 rooms. The two islands only feature hotels and transportation infrastructure, with a complete lack of other services facilities. Cuban tourist enclaves are characterized by a very high degree of spatial isolation that is achieved by island locations as well as a complete lack of non-tourist infrastructure. There are no permanent settlements, with hotels and transportation infrastructure being the only type of development available on site. Two enclaves (Cayo Largo and Cayo Santa Maria) include shopping centers (known as *puebla*) whose main purpose is to link hotel complexes and form urbanized zones between them as well as to provide services not provided by hotels (Pueblo la Estrella, Pueblo Las Dunas).

Tourist enclave operations generate environmental pressure due to everyday processes. The delicate ecological balance of the island ecosystems may become disturbed by expansive tourist centers (García *et al.*, 2015; Gutiérrez, 2015; Muñuzuri *et al.*, 2012). The Cuban government has noted environmental problems encountered on small islands, including soil degradation, deforestation, inland water pollution and threats to biodiversity. Tourist islands are likely to be deforested by developers building new tourist infrastructure, while mangroves are used to produce fuel, as is peat. Natural biodiversity in the region is threatened by the introduction of exotic decorative plants. The islands are formed of sand bars and face a constant shortage of drinking water, which is obtained from wells on the main island and piped to everywhere else.

The problem of solid waste is also becoming an issue on islands with hotels, as every tourist generates a daily average of 1.5 kg of garbage. Environmental problems also affect the waters in the area, which experience increasing wastewater runoff and poor water exchange in basins surrounding the studied islands (Jardines del Rey) separated by causeways. These barriers also make it difficult for marine animals to migrate between adjacent bodies of water. Excessive concentration of tourist traffic has led to destruction in coastal areas including coral reefs and beaches. It may soon turn out that tourism has initiated the process of environmental self-destruction across the studied archipelago.

New laws and regulations have been enacted in order to protect coastal areas from the negative effects of tourism development. One of the more important regulations concerns the free use of the coastal zone or the land area between 20 and 40 m (depending on coast type) from the shoreline. On the other hand, sea areas between 100 and 200 m from the shoreline must remain free of permanent shoreline infrastructure such as piers and docks. Yet another step designed to stop uncontrolled development in tourist areas consists of a national spatial management plan whose main purpose is to designate tourist areas and to plan for new tourist areas. Tourism development in Cuba is supposed to be limited to 16 regions – most of which are coastal regions.

In addition to the large increase in the number of hotel beds, the area is also affected by gradual change in the quality of services. The internet pages and catalogs of the largest tour operators serving this destination were used along with fieldwork to draw the following conclusions:

- Cayo Coco and Cayo Guillermo used to be treated as secondary tourist regions and would normally be outpaced by the Varadero region (Havana), which features a more diversified tourism infrastructure as well as greater potential in terms of material and non-material culture.
- Cayo Coco and Cayo Guillermo are characterized by very good external transportation options (Cayo Coco Airport) and less optimal local transportation options, which are often a problem in the organization of educational trips to the largest area attractions (Havana – 510 km, Trinidad – 250 km).
- Despite the availability of some roads, geographic isolation limits recreational activity to the island itself, while a lack of additional infrastructure such as shopping centers, sports centers, fitness centers and recreational parks limits recreational activity to the grounds of the tourist centers themselves.
- The lack of permanent settlements results in the lack of a local population and the benefits that normally accrue in such cases. A local population is an important element of the local job market and yields local cultural potential. The absence thereof decreases the attractiveness of the area to tourists.

The above features strongly affect the price of the offered tourist packages and the length of the average tourist stay. The vacation offerings of travel agencies in Canada, being the main country generating tourist traffic in Cuba, were examined in the study. Research has shown that a vacation in Cayo Coco and Cayo Guillermo is less expensive than one in Varadero by an average of 100 Canadian dollars per week. The packages compared in this case were for analogous time periods and similar levels of service quality. One package offered for Cayo Coco and Cayo Guillermo consists of a three-day weekend – a rarity in the case of Varadero, where a 'standard' stay is three weeks or more. European tour operators usually offer three vacation durations for most regions in Cuba: one week, 10 days and two weeks. European travel offerings tend to focus on Varadero, but stays at Cayo Coco and Cayo Guillermo are not priced differently.

The above analyses suggest that the Cayo Coco and Cayo Guillermo tourist region is a typical recreational tourism region with limited opportunities for educational excursions. This feature of the region results in a relatively low occupancy rate (low season – 47%, high season – 75%) compared with the Varadero region. In order to change this situation, it was decided to build exclusive hotels on the islands that will attract customers seeking ultra-luxury accommodations. One example of this is the Grand Centara Beach Resort Cayo Guillermo, which is still under construction in the western part of the island and will receive its first guests in 2018. This will be a hotel with 245 spacious apartments that feature private pools in some cases. The hotel will also offer a wide array of spa services. The number of such luxury hotels is supposed to increase to six in the coming years. The construction of this type of hotel is designed to change the nature of the tourist offering of this destination in order to target a new group of tourists – exclusive tourists. This newer group of tourists is slated to generate much more revenue per day than is the case at present.

4 Discussion

The above research results suggest the direction in which tourism should evolve on small islands such as Cayo Coco and Cayo Guillermo. The following question is worth asking: Does it make sense to keep developing tourist enclaves?

The scientific community has been debating the purpose of tourist enclaves for more than a decade. There are arguments for and against tourist enclaves and these deal with a wide array of issues, from the natural environment to economic growth and social development. It is difficult to unequivocally determine the weight of positive and negative arguments in this key discussion (Núñez & Fernández, 2011). The impact produced by tourist enclaves is difficult to assess with respect to all components of the geographic environment. It is often the case that planned, controlled and formalized tourist centers are less burdensome on the natural environment than the uncontrolled growth of small tourist facilities that are unable to fulfill environmental requirements and tend to escape regulations by operating illegally.

The low-level outflow of tourism-related profits in the service sector and the linkages between the tourism sector and other sectors of the national economy will always remain a matter for discussion. However, this is a consequence of the need to acquire foreign investors in the face of meager local resources. Arguments against tourist enclaves include a change in the way people travel by moving away from the 3S model of recreational tourism and towards other forms of tourism including cultural, sports-oriented and ecological tourism. This slow transition is observed in many countries, but it often only applies to the most conscious and best educated tourists and not to all tourists. Perhaps many tourists still wish to experience the feeling of luxury during an all-inclusive vacation in an exotic part of the world.

Changes introduced by the government of Barack Obama and later modified by the government of Donald Trump have led to a situation where US citizens may travel to Cuba under certain conditions (Apollo & Rettinger, 2019). Americans can most easily enter Cuba as recreational tourists who choose to travel as part of an organized tour guided by licensed operators. This change led to an increase in demand for high-quality tourist centers offering all-inclusive services. Hence, the government of Cuba made the decision, along with foreign investors, to increase the number of hotel beds on the island – and this is only made possible by constructing new hotels.

Currently the Cayo Coco and Cayo Guillermo region is visited by 320,000–350,000 tourists per year, who may choose from about 5000 hotel beds managed by 3700 employees daily. It is estimated that the region will have about 765,000 tourists per year by 2020 and the number of hotel beds will increase to 12,400, while 8000 workers will find employment in these hotels and other facilities (Muñuzuri *et al.*, 2012). The next few years will see rapid construction activity on the islands, which will reduce the quality of the tourist experience for current tourists and will increase tourist traffic in the future.

It appears that all actions in the realm of tourism development are supply based. It is assumed that all actions associated with tourism development in Cayo Coco and Cayo Guillermo will naturally lead to an increase in tourist expenditures and consequently an increase in sector profits. The difficult economic situation in Cuba forces a focus on increasing revenues from tourism. This aspect of tourism is considered to be a priority and social and environmental issues are considered secondary.

Future spatial development should be based on a broad analysis of key determinants – particularly environmental determinants, which are a fundamental issue as recreational tourism is largely determined by the quality of the local natural environment. The loss of local environmental attractions leads to a loss of recreational tourists. One additional key argument in the formulation of a sustainable development strategy is the fact that the two islands are small and difficult in terms of maintaining environmental balance. A plan to build such large recreational facilities requires proper procedures and practices on the part of investors and hotel guests in the areas of pollution, water management and waste disposal. All-inclusive resorts use a lot of water and produce a lot of waste. These types of facilities rely on the environmental potential of a region, but also tend to put a lot of pressure on the local natural environment (Whittle *et al.*, 2003).

In terms of transportation options, tourist regions located on islands are somewhat isolated. In particular, this is true of Jardines del Rey – an archipelago of small islands at a distance of about 30 km from the Cuban mainland, which is home to most attractions in Cuba. The accessibility of the region is based on two pillars: an airport on Cayo Coco Island with a capacity of 600 passengers per hour as well as causeways that link the archipelago with the mainland (Muñuzuri *et al.*, 2012).

Regarding local economic development, the region's economy should not be based solely on tourism and should strive for diversification (Rettinger & Apollo, 2017). However, this tourist region was founded in an area with no permanent population. In addition, the Cuban government wishes to isolate tourists from the Cuban community in general and does not permit small firms to establish a foothold in the tourism sector in the region.

5 Conclusion

Further development of tourist enclaves featuring highly ranked all-inclusive hotels is not consistent with world trends in tourism, which are based on non-material attractions and intangible assets such as feelings, experiences, social atmosphere and overall well-being. Changes in the tourism sector are driven, first and foremost, by changes in the tourist community itself. The modern tourist is wealthier, more demanding, better educated, more worldly and seeking to engage with local communities in tourist areas. In this context, does it make sense to build more tourist enclaves in the way in which the Cuban government is doing? The answer lies in the demand for such enclaves. Will new tourists keep coming? This requires a broad study of the motivation behind travel and tourism and recreation in isolated areas with limited access to local populations. One way to study the tourist enclave concept in Cuba is to examine patterns associated with Canadian tourists who often treat Cuba as a 'weekend' destination. Cuba is very popular with Canadians who value its relatively low prices, very good transportation options, comfortable accommodation and overall security at tourist destinations. All of these key elements form what may be described as the visitor space of Cuba.

It is expected that tourist enclaves will continue to expand rapidly in the following six key tourist areas of Cuba: Cayo Jutía and Levisa in Pinar del Río province; Cayo Largo del Sur in Isla de la Juventud province; Cayo Santa María, Las Brujas and Ensenachos in Villa Clara province; Cayo Coco, Guillermo and Romano (Jardines del Rey) in Ciego de Ávila province; Jardines de la Reina and Sabinal Key in Camagüey province; and Cayo Saetía in Holguín province. Tourist enclaves will continue to be built on islands or in spatially isolated geographic areas.

Acknowledgement

This chapter has been previously published, as Rettinger, R. (2019) Cuban tourist enclaves: A critical approach. In *6th SWS International Scientific Conferences on Social Sciences 2019* (pp. 65–72). doi:10.5593/SWS.ISCSS.2019.2/S04.009; ISBN 978-619-7408-92-8; ISSN 2682-9959. Permission for republication has been gained from SGEM GeoConference Team.

References

Apollo, M. and Rettinger, R. (2019) Mountaineering in Cuba: Improvement of true accessibility as an opportunity for regional development of communities outside the tourism enclaves. *Current Issues in Tourism* 22 (15), 1797–1804.

Caribbean Tourism Organization (2016) *Caribbean Tourism Statistical Report*. Barbados.

Cerviño, J. and Cubillo, J.M. (2005) Hotel and tourism development in Cuba: Opportunities, management challenges, and future trends. *Cornell Hotel and Restaurant Administration Quarterly* 46 (2), 223–246.

Derek, M. (2007) Od wioski rybackiej do turystycznej enklawy? Turystyka a rozwój lokalny na przykładzie Mikołajek. *Studia Regionalne i Lokalne* 28 (2), 112–127.

García, M.E.B., Rodríguez, M.C.F. and Peña, L.B. (2015) 'Jardines del Rey': An integrated sustainable management of a Cuban touristic destiny. In A. Panosso Netto and L.G.G. Trigo (eds) *Tourism in Latin America: Cases of Success* (pp. 143–157). Cham: Springer.

Gutiérrez, L.R. (2015) The environmental effects of tourism architecture on island ecosystem in Cayo Guillermo, Cuba. *Journal of Environmental Protection* 6 (9), 1057.

Issa, J. and Jayawardena, C. (2003) The 'all-inclusive' concept in the Caribbean. *International Journal of Contemporary Hospitality Management* 15 (3), 167–171.

Kowalczyk, A. and Derek, M. (2010) *Zagospodarowanie turystyczne*. Warsaw: Wydawnictwo Naukowe PWN.

Lee, C.F. and King, B. (2006) Assessing destination competitiveness: An application to the hot springs tourism sector. *Tourism and Hospitality Planning & Development* 3 (3), 179–197.

Matthews, H.G. (1977) Radicals and third world tourism: A Caribbean focus. *Annals of Tourism Research* 5, 20–29.

Muñuzuri, J., Onieva, L., Cortés, P. and Guadix, J. (2012) Design of a transportation system for workers in a Cuban resort area. *WIT Transactions on Ecology and the Environment* 166, 15–26.

Núñez, G.G. and Fernández, R.O. (2011) Cuba y el turismo norteamericano. Analisis de potencialidades y de impactos en la región caribeña. In *Political Economy of Change in Cuba* (pp. 141–199). New York: Bildner Center for Western Hemisphere Studies. See https://thecubaneconomy.com/wp-content/uploads/2011/09/2011-09-21-PEchangeCuba-2.pdf#page=131

Padilla, A. and McElroy, J.L. (2007) Cuba and Caribbean tourism after Castro. *Annals of Tourism Research* 34 (3), 649–672.
Rettinger, R. and Apollo, M. (2017) Enriching tour-operators' offer to go beyond sea, sun and sand: The case of the Dominican Republic. In R. Efe and M. Öztürk (eds) *Contemporary Studies in Environment and Tourism* (pp. 60–81). Newcastle upon Tyne: Cambridge Scholars Publishing.
Rettinger, R. and Wójtowicz, M. (2014) Regional differences in the development of tourism in Cuba. In M. Wójtowicz and A. Winiarczyk-Raźniak (eds) *Environmental and Socio-economic Transformations in Developing Areas as the Effect of Globalization* (pp. 136–154). Krakow: Wydawnictwo Naukowe UP.
Theuns, H.L. (2011) Analiza strukturalna rozmiarów rzeczywistych i potencjalnych wycieków/powiązań w gospodarce turystycznej Karaibów. *Folia Turistica* 25 (2), 123–152.
Whittle, D.J., Lindeman, K.C. and Tripp, J.T.B. (2003) International tourism and protection of Cuba's coastal and marine environments. *Tulane Environmental Law Journal* 16, 1–57.

15 The Impact of Environmental Factors on Profit Efficiency in Rice Production in Northern Vietnam

Hoang Van Long and Mitsuyasu Yabe

Environmental factors affect the productivity and efficiency of agricultural production, resulting in a change to profit efficiency. This chapter seeks to estimate the impact of environmental factors on the profitability of rice farming in the Red River Delta of Vietnam. The dataset was compiled from personal interviews with 349 farmers selected from more than 1000 rice farming households in four different regions. Both OLS and MLE translog profit functions were used in this study, with five production inputs and four environmental factors. The estimation of the stochastic profit frontier with a two-stage approach was used to measure profitability. The results show that the profit efficiency was about 75% on average and that various environmental factors influence profit efficiency significantly in addition to farm-specific characteristics. Plant disease, soil fertility, irrigation management and water pollution are the four environmental factors that cause profit loss in rice production. The results of the study indicate that farmers should reduce both their household size and the number of farm plots, apply a row seeding technique and improve environmental factors to obtain high-profit efficiency. Special consideration should be given to irrigation infrastructure and water quality improvement.

1 Introduction

Environmental degradation is a consequence of economic and industrial development in developing countries, particularly Vietnam. Water pollution has also become a major concern as pollutants from industrial activities contaminate rice fields and affect the country's rice production.

When negative environmental factors are omitted, both the technical efficiency and the profitability of rice production improve (Rahman & Hasan, 2011). Therefore, environmental factors such as soil quality, irrigation management, plant disease and water pollution were primary considerations in this study. Farmers may combat environmentally constraining factors by allocating more labour and adding more chemical fertiliser to the input bundle of their production (Sherlund *et al.*, 2002). These activities may lower productivity and increase inefficiency. As a result, advanced technologies, such as using machinery and new seeds and fertilisers, become major factors to improve productivity.

There were four objectives of this research: (1) to determine the effects of environmental factors on the profit efficiency of rice production; (2) to estimate the profit loss of rice production due to environmental factors; (3) to provide recommendations to policymakers on how to sustain rice production; and (4) to contribute to the literature on the methodological development of estimating the impacts of environmental factors on rice production.

The next section of this chapter provides a review of the literature related to the estimation of efficiency with special consideration given to profit functions. In the third section, details of the research areas are explained, including the data from the household survey. The framework for analysis and the econometric specification are presented in the fourth section. The fifth section is the major component, presenting results and discussions from model estimation. The final section offers conclusions and policy implications.

2 Rice Production in Vietnam: Growth, Constraints and Perspectives

Vietnam is the second highest rice exporter in the world. Rice production is important to the Vietnamese economy in general and to its agriculture in particular. It occupies a high share in the country's Gross Domestic Product (GDP), contributing about 20.4% of its GDP in 2006 (WB, 2008). The rice export value increased 10-fold from US$275 million in 1990 to US$2.9 billion in 2010. It then increased to exceed US$3 billion in 2010 and declined to US$2.1 billion in 2016 (Vietnam Food Association, 2018). Constraints that affect the rice production sector include the lack of a Vietnamese rice brand on the world market, the reduction of land available for rice production, and productivity and profitability becoming more competitive with other plant varieties and environmental pollution.

In terms of production area, the total rice growing area of the country increased from 6.30 million hectares (ha) in 1990 to 7.67 million ha in 2000. It then declined to 4.20 million ha in 2010. This decrease of more than 100,000 ha annually resulted from rapid economic expansion and industrialisation. Consequently, land for rice production became narrower and less fertile. On the other hand, land fragmentation also places a constraint on crop productivity and machinery application in cultivation. The whole country has about 75 million parcels of land for rice production, and one household typically holds about seven to eight plots (Hung *et al.*, 2007).

Furthermore, uncultivated and low productivity areas have appeared due to pollution from the expansion of industrial production and traditional craft production villages (Dung, 2009; Mahanty & Dang, 2013). Therefore, the challenges that have arisen within the rice production and export sector mean that it will be more difficult to keep increasing the export value and lowering the impact from environmental pollution factors, such as soil pollution and water irrigation quality.

Policymakers are trying to improve the rice production sector in both quantity and quality. It is important not only for rice farmers but also for the authorities at all levels to determine the factors that affect production in order to improve understanding of and recommendations for minimising inefficiencies in productivity and profitability. The research focused on the analysis of factors affecting profitability using the production function approach. The next section provides the detailed framework of the approach and the econometrics model results.

3 Profitability Analysis Framework

Understanding the profitability of rice production has become a major concern not only for farmers but also for policymakers. In particular, there is a need to understand how to shift the profit of rice production to the profit frontier under the condition of limited land. Industrial development and the development of handicraft production are the major drivers causing the worsening quality of water for rice production. Some studies have been conducted in relation to efficiency measurements of rice production, such as technical efficiency (Khai & Yabe, 2011; Vu, 2008) and productivity (Long & Yabe, 2011). There are three research questions that arise in the studies. Firstly, how can the profit elasticity of rice production factors be estimated? Secondly, is there any loss due to inefficient use of production factors, socioeconomic characteristics of farming households and environmental factors? Thirdly, what is the profit inefficiency of profit function? However, research related to profit efficiency is limited in the literature. Therefore, this study aimed to estimate the three dimensions of profitability of rice production, namely profit elasticity, profit loss and profit inefficiency. In addition, the environmental attributes affecting the profitability of rice production were determined. Economic efficiency is classified into two components: technical efficiency and allocative efficiency (Farrell, 1957). The profit function combines both technical and allocative concepts in a profit relationship, and any errors in production decisions are translated into lower revenue (Ali *et al.*, 1994) and, hence, low profit efficiency.

$$\ln\pi' = \alpha_0 + \sum_{i=1}^{4} \alpha_i \ln P_i' + \frac{1}{2}\sum_{i=1}^{4}\sum_{h=1}^{4} \gamma_{ih} \ln P_i' \ln P_h' + \sum_{i=1}^{4}\sum_{k=1}^{4} \delta_{ik} \ln P_i' \ln Z_k$$
$$+ \sum_{k=1}^{4} \beta_k \ln Z_k + \frac{1}{2}\sum_{k=1}^{4}\sum_{j=1}^{4} \theta_{kj} \ln Z_k \ln Z_j + \varepsilon$$

where π' equals restricted profit (total revenue less total cost of variable inputs) normalised by the price of output (P_y); P_i' is the price of the *i*th input (P_i) normalised by the output price (P_y); *i* equals (1) fertiliser price (kg^{-1}), (2) labour wage (day^{-1}), (3) animal power price (animal pair-days^{-1}), (4) seed price (kg^{-1}); Z_k equals the quantity of fixed input; *k* equals (1) area under wheat (ha^{-1}), (2) irrigation (farm^{-1}), (3) land suitability (index number farm^{-1}), (4) other environmental constraints (index number farm^{-1}); ε refers to random error; ln is natural logarithm; α_0, α_i, γ_{ih}, β_k, δ_{ik} and θ_{kj} are the parameters to be estimated.

The profit frontier approach is defined as:

$$\pi_i' = f(P_i, Z_i)\exp(\xi_i)$$

where π_i' is the normalised profit of farm *i*; P_i' is the normalised input prices measured by dividing profit and input prices for output prices; Z_k is the fixed inputs such as land and capital.

This function can be estimated by OLS (Ordinary Least Squares) of MLE (Maximum Likelihood Estimation) (Ali & Flinn, 1989). The OLS approach of profit function is written as

$$\ln\pi_i' = \alpha + \sum_{j=1}^{j} \alpha_j \ln P_j' + \alpha_L \ln Z_k + \xi_i$$

The translog profit function approach was used by Aigner *et al.* (1977), Meeusen and Van Den Broeck (1977) and Ali and Flinn (1989). The translog frontier form can be written as

$$\ln\pi_i' = \alpha_0 + \sum_{j=1}^{j} \alpha_j \ln P_j' + \frac{1}{2}\alpha_j\left(\ln P_j'\right)^2 + \alpha_L \ln Z_k + \frac{1}{2}\alpha_Z\left(\ln Z_k\right)^2$$
$$+ \sum_{j=1}^{j}\sum_{k=1}^{k} \alpha_{jZ} \ln P_j' \ln Z_k + \xi_0$$

This study used both forms of profit frontier to compare the coefficients between different approaches to estimation. Production inefficiency

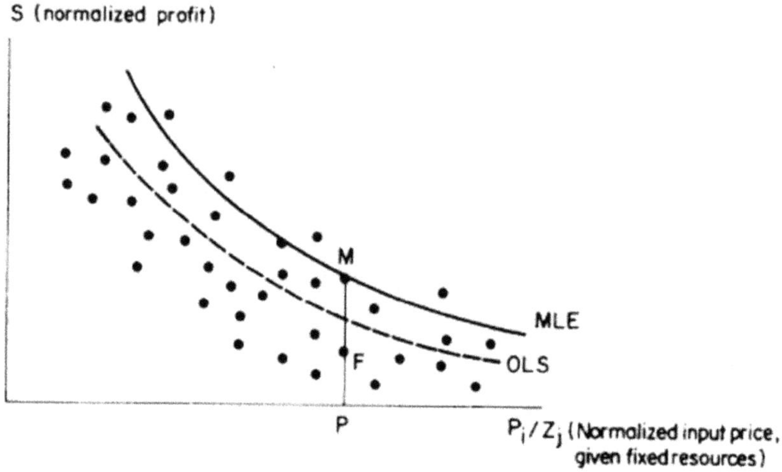

Figure 15.1 Normalised profit and input price
Source: Ali and Flinn (1989).

is measured by three components: technical, allocative and scale inefficiency. Error terms are $\xi_i = v_i - u_i$.

Production/profit efficiency of the individual farm i is defined as

$$PE = E\left[\exp\left(-u_i\right)\middle|\xi_i\right] = e\left[\exp\left(-\delta_0 - \sum_{d=1}^{D}\delta_d W_d\right)\middle|\xi_i\right]$$

Besides the farm and household characteristics, the environmental factors/constraints are used to estimate efficiency in areas such as irrigation, land suitability, insects and pests, weed infestation, weather variation (drought and storm) and soil fertility status (Rahman & Hasan, 2011). In this study, four environmental factor variables (W) were selected: irrigation, plant disease, soil fertility and water quality.

The estimation of profit efficiency was determined by two different approaches. The OLS estimates the average profit value while the MLE estimates the profit frontier (Ali & Flinn, 1989). Using the results of the MLE approach, the profit loss could be derived by dividing the profit of the individual farms for profit efficiency (Figure 15.1).

4 Research Area, Data and Econometric Specification

4.1 Research area and data

The Red River Delta is at the heart of rice production for all of North Vietnam. Although rice productivity has been gradually increasing in recent years due to the adoption of advanced technologies, production still faces challenges from issues such as land fragmentation, soil degradation and water pollution. One reason for this situation is the development of industrial and handicraft production. Many industrial parks and production zones have been established near the rice fields, and handicraft villages have also increased rapidly. Recently, the total number of craft villages has risen dramatically to 2790 across the country, and half of them are located in the Red River region. This development has led to increased pollution (Quynh *et al.*, 2003). Water pollution is at an alarming level, with about 90% of craft villages violating environmental standards (MOLISA, 2009).

Rice production in Bac Ninh province still plays an important role in the economic development of the region, with some special rice species that are well known for their flavour. The province has the highest number of craft villages ($n = 61$) in the country. Water pollution from these craft villages causes some tens of hectares of land in the region to be uncultivable and farm households to become landless.

The research was conducted in Bac Ninh province (Figure 15.2) and one part of Hanoi, the capital, where rice production still dominates the local economy. These areas were selected to study the effect of water pollution on rice production in parallel with other environmental factors. Four communities were selected for the survey, namely Phong Khe, Da Ton, Phu Dong and Ninh Hiep. Phong Khe is located in Bac Ninh Province, while the three remaining communities are in nearby areas of Hanoi. Phong Khe and Da Ton are situated in polluted areas, while the remaining two communities are in non-polluted areas.

Phong Khe is located near a paper recycling village and Da Ton is near Sai Dong Industrial Park. In Sai Dong, pollutants from electronics companies are discharged into the nearby river which flows to an irrigation canal used by farmers. Phong Khe is in Bac Ninh province, which is famous for its handicraft production. Paper recycling is one of the major local industries in this community. The industry uses a lot of 'cleaning' and 'colourful' chemicals. A significant amount of wastewater is discharged directly into the river without treatment. Therefore, almost all of the local rice paddy areas are contaminated. Ninh Hiep is less polluted because it is not near any production areas and irrigation is from the big Red River. Finally, Phu Dong is located at the last stage of a branch of the Red River. Water pollution in this area is relatively low. It comes mainly from household waste, which does not affect rice production significantly.

A survey of rice farmers was conducted in these four communities in two provinces of the Red River Delta. A total of 369 rice farmers were interviewed using a structured questionnaire. The survey was conducted in August 2010 by a group of enumerators who received basic training. A pre-test was conducted to revise the questionnaire before the formal survey. After computing the profit, 20 households were deleted from the sample. In the end, the sample used for profit function estimation was 349 households. The 20 households were dropped from the dataset because of missing variables for profit function and minus values for profit.

Table 15.1 shows the descriptive statistics of all variables in the dataset for the profit function: rice output (kg), rice yield (kg/sao), land (sao), the

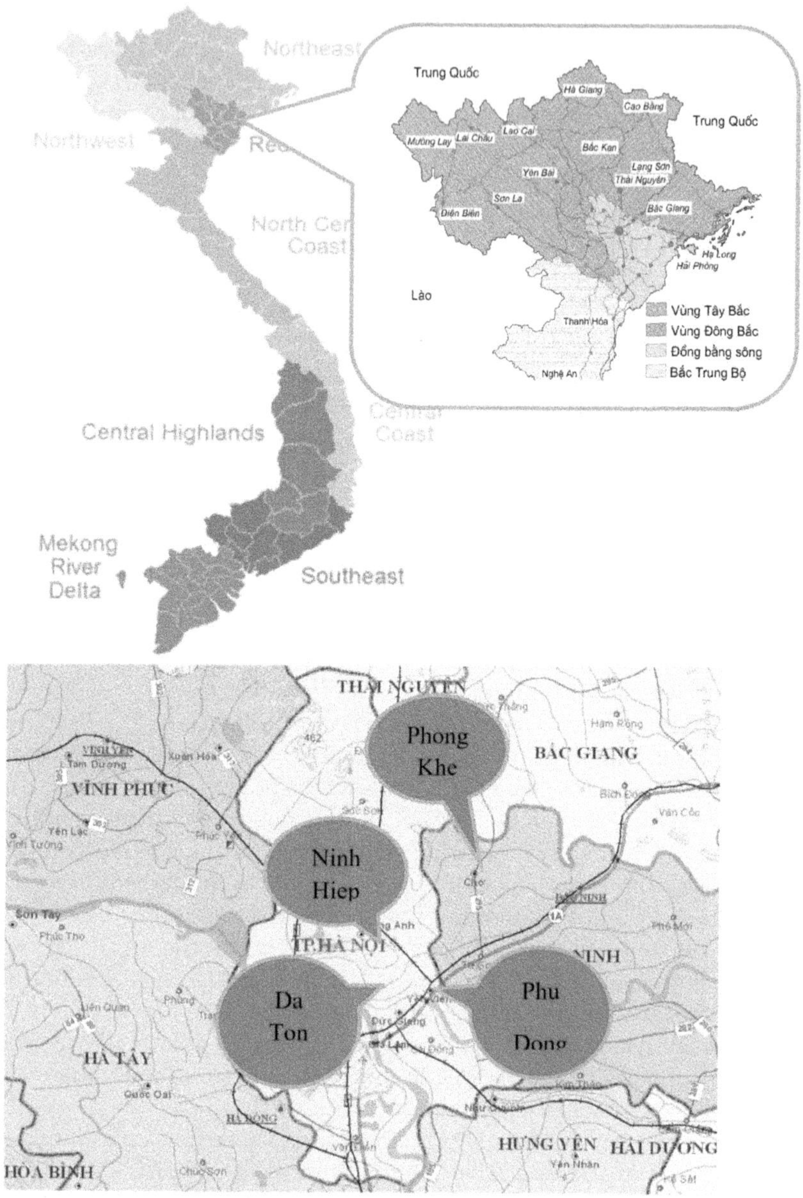

Figure 15.2 Map of research sites

price of fertiliser (VND '000/kg), pesticides (VND '000/100 ml) and labour wages (VND '000/working day per man), land input (sao per farm; 1 sao = 360 m^2) and capital (VND '000/farm). All of these variables were calculated on a yearly basis. Only five inputs were used for model estimation.

The environmental factors – soil quality, irrigation, disease and water pollution – are defined as dummy variables. Managerial variables are the

Table 15.1 Description of variables

Descriptions	**Measure**	**Mean**	**SD**	**Min**	**Max**
Output and input:					
Rice output	kg per year per farm	1552.829	1198.260	90	10400
Rice out price (PO)	VND '000 per kg	5.335	0.296	4.250	6.800
Rice yield	kg per sao per annum	178.010	44.496	62	700
Fertiliser price (F)	VND '000 per kg	5.806	1.895	1	20
Pesticide price (PE)	VND '000 per 100 ml	10.183	9.296	0.9	75
Labour wages (WA)	VND '000 per working day	73.272	11.943	33	110
Land area cultivated (L)	Sao (1 Sao = 360 m^2)	8.472	5.446	0.5	40
Capital (C) – rental cost for land preparation, harvesting and transportation services	VND '000 per farm per annum	713.301	584.748	0	4484
Environmental factors:					
Soil quality	Dummy (1 = good; 0 = bad)	0.580	0.494	0	1
Irrigation	Dummy (1 = yes; 0 = no)	0.954	0.210	0	1
Diseases	Dummy (1 = yes; 0 = no)	0.995	0.074	0	1
Water pollution	Dummy (1 = yes; 0 = no)	0.534	0.500	0	1
Managerial variables:					
Age of household head	Years	48.125	9.700	26	75
Male household head	Dummy (1 = male; 0 = otherwise)	0.496	0.501	0	1
Household size	Number of members	4.792	1.515	2	10
Education of household head	Completed years of schooling	6.336	3.110	0	12
Experience in rice farming of household head	Years of rice growing	27.756	12.005	0	57
Family labour ratio	Ratio of family labour and household size	0.395	0.297	0	1
Number of rice plots	Number of plots of rice fields that household cultivates	4.016	2.317	1	17
Monocropping	Dummy (1 = yes; 0 = no)	0.081	0.273	0	1
Row seeding	Dummy (1 = row seeding; 0 = broadcasting or other)	0.154	0.362	0	1
Credit	Dummy (1 = borrow money for rice production; 0 = don't borrow)	0.012	0.137	0	1
Total number of observations		**349**			

Note: US$1 equivalent to Vietnamese dong (VND) 20,000 in 2010.

age of the household head, gender of the household head, education of the household head and the family labour ratio. In addition, the number of rice plots, monocropping, adoption of row seeding technology and access to credit are included.

4.2 Econometric specification

Measuring efficiency, the ability to produce a given level of output at the lowest cost, was first done by Farrell (1957). Ali and Flinn (1989) estimated profit efficiency by comparing the OLS with the MLE approach to show the profitability of rice production in Pakistan. Rahman (2003) estimated rice profit efficiency using the translog function, adding the characteristics of farmers to the inefficiency effects. Kolawole (2006) estimated the profit function, adding a constant to profit function to obtain the positive values and used the Cobb–Douglas function to estimate the profit frontier.

The standard profit function assumes that markets for outputs and inputs are perfectly competitive.

$$\ln(\pi+\theta)=\ln f(P,W)+(u-v)$$

where θ is a constant added to the profit (π) of each firm in order to obtain positive values (Kolawole, 2006).

The farm profit was measured in terms of Gross Margin (GM), which equals the difference between the Total Revenue (TR) and the Total Variable Cost (TVC):

$$GM(\pi) = \Sigma(TR - TVC) = \Sigma(QP - WX_i)$$

To estimate the impacts of environmental factors on profit efficiency, first, the stochastic profit function is defined as

$$\pi_i' = f(P_i', Z_i)\exp(\xi_i)$$

where π_i' is a normalised profit of the ith farmer defined as gross revenue less variable cost, divided by farm output price; P_i' is the vector of variable input prices faced by the ith farmer divided by output price (normalised price); Z_i is the vector of fixed factor of ith farmer; ξ_i is an error term; and i equals 1,..., i is the number of farms in the sample.

The model was first developed by Sidhu and Baante (1981) and applied by Rahman (2003) and Hyuha *et al.* (2007).

$$\ln\pi_0' = \alpha_0 + \sum_{j=1}^{6} \ln P_j' + \frac{1}{2}\sum_{j=1}^{6}\sum_{k=1}^{6} \tau_{jk} \ln P_j' \ln P_k' + \sum_{j=1}^{6}\sum_{l=1}^{2} \varnothing_{jl} \ln P_j' \ln Z_l$$

$$+\sum_{l=1}^{2} \beta_l \ln Z_l + \frac{1}{2}\sum_{l=1}^{2}\sum_{t=1}^{2} \varphi_{lt} \ln Z_l \ln Z_t + u_0 - v_0$$

where P_j and P_k are the (1) price of seed, (2) price of fertiliser, (3) price of organic fertiliser, (4) price of pesticides, (5) price of herbicides and (6) labour wages ($j, k = 1$–6, respectively). Z_l and Z_t are land area and capital of each farm ($l, t = 1$–2, respectively).

$$u_0 = \delta_0 + \sum_{d=1}^{13} \delta_d W_d + \omega_0$$

where W are the variables representing six socioeconomic characteristics: (1) age of head of household, (2) male head of household, (3) education of household head (number of completed year of schooling), (4) household size, (5) family labour ratio, (6) rice plots (the number of plots that households use to cultivate rice); three farm and technical characteristics: (1) monocropping (dummy for households that cultivate one crop per year), (2) row seeding technique, (3) taking a loan for rice production; and four environmental factors: (1) dummy for soil quality, (2) dummy for diseases, (3) dummy for irrigation use, (4) dummy for water pollution. These variables are the determinants to explain profit inefficiency.

5 Results and Discussion

5.1 Profit efficiency

Comparing the results of OLS and MLE shows a small change in coefficients when using different methods to estimate the profit frontier. In addition, the OLS estimation result shows that 69% of the dependent variable (profit) can be explained by independent variables (production inputs). This indicates that the production profit is close to the profit frontier.

Fertiliser price shows a positive impact on profit. The reason is that chemical fertilisers are varied. The price was therefore estimated by the average price of all fertilisers per kilogram, showing that a higher price of fertiliser means a higher quality of fertiliser; therefore, it is rational that it will increase profit. In this study, a higher priced fertiliser means a higher quality of fertiliser, indirectly increasing the yield. An increase in the labour wage would indicate farmers using more hired labour than home labour, which also increases the yield. The capital lowers the profit because capital is considered to be input and it is better if the farmer can rent the capital.

The normalised price variables (P_i') were calculated by dividing the price of input prices (i) with output price (PO). The five input prices (i) in the function are F (fertiliser price), PE (pesticide price), WA (labour wages), L (land area cultivated) and C (capital).

The model estimations for profit function using the two methods (OLS and MLE) are presented in Table 15.2. It shows that the results are similar in these two estimation methods.

Table 15.2 Model estimations for profit function

Variables	Parameters	OLS estimation Coefficients	OLS estimation *t*-ratio	MLE (frontier estimation) Coefficients	MLE (frontier estimation) *z*-value
Profit function					
constant	α_0	6.029***	6.85	6.324***	8.14
$\ln P'_F$	α_1	0.228	0.28	0.784	1.1
$\ln P'_{PE}$	α_2	0.163	0.38	0.344	0.9
$\ln P'_{WA}$	α_3	−0.615	−1.14	−0.509	−1.09
½ $\ln P'_F \ln P'_F$	α_4	−0.475**	−2.44	−0.451***	−2.88
½ $\ln P'_{PE} \ln P'_{PE}$	α_5	−0.042	−0.7	−0.064	−1.19
½ $\ln P'_{WA} \ln P'_{WA}$	α_6	0.523**	2.28	0.466***	2.44
$\ln P'_F \ln P'_{PE}$	α_7	−0.114	−1.23	−0.025	−0.31
$\ln E'_F \ln P'_{WA}$	α_8	0.244	0.96	0.210	0.99
$\ln P'_P \ln P'_{WA}$	α_9	−0.019	−0.12	−0.031	−0.24
$\ln C$	α_{10}	0.138	0.63	0.045	0.24
½ $\ln C \ln C$	α_{11}	−0.028*	−1.61	−0.003	−0.18
$\ln P'_F \ln C$	α_{12}	−0.128	−1.1	−0.236**	−2.3
$\ln P'_P \ln C$	α_{13}	−0.024	−0.44	−0.0349	−1.14
$\ln P'_W \ln C$	α_{14}	−0.028	−0.35	−0.027	−0.39
$\ln P'_F \ln L$	α_{15}	0.037	0.27	0.115	0.98
$\ln P'_P \ln L$	α_{16}	0.005	0.08	0.051	0.53
$\ln P'_W \ln L$	α_{17}	0.025	0.16	0.036	0.27
$\ln L$	α_{18}	0.013**	0.03	−0.0x10	−0.03
½ $\ln L \ln L$	α_{19}	0.353***	7.34	0.341***	7.58
R-squared		0.690			
Number of observations		**349**		**349**	

Notes: *** Significant at 1% level ($p < 0.01$); ** significant at 5% level ($p < 0.05$); * significant at 10% level ($p < 0.10$).

The variance parameters are only available in the MLE estimation. They show that the error terms (v and u) are significant in estimations with high z-values (Table 15.3). Land area positively affects profit. This means that the margin profit increases when the area of cultivated land increases. Also, this shows the economies of scale in rice production.

The parameter estimates of the profit function model were used to estimate the elasticities (see Table 15.4). The elasticity estimates indicate that rice farmers are responsive to changes in prices. Land area dominates the profit share. Increasing the land area by 1% increases the profit by 1.2%; if the price of rice increases by 1%, the profit increases by 0.7%. If the price of fertiliser increases by 1%, the profit increases by 0.07%. If the

Table 15.3 Variance parameters

Variance parameters		Parameters	z-value
$\ln\sigma v^2$		−4.000	−12.66
$\ln\sigma u^2$		−1.850	−12.42
$\sigma^2 = \sigma u^2 + \sigma v^2$	σ^2	0.170	
$\lambda = \sigma u/\sigma v$	Λ	3.313	
Number of observations		**349**	

Table 15.4 Estimation of profit elasticities

Prices and fixed inputs	Profit elasticity
Rice price	0.690
Fertiliser price	0.072
Pesticides price	−0.086
Labour wages	0.956
Land area	1.257
Capital	−0.090

price of pesticides increases by 1%, the profit reduces by 0.08%. If the labour wages increase by 1%, the profit increases by 0.9%. If the capital increases by 1%, the profit decreases by 0.09%.

5.2 Profit inefficiency and profit loss

Profit inefficiency was estimated by OLS regression estimation of error terms from the profit function. (Table 15.5) using farm and household characteristics and environmental factors.

A higher age of the household head increases profit inefficiency. Young head farmers can work more effectively than their older counterparts. Household size affects profit inefficiency because households with more members cannot utilise home labour more effectively than smaller households. Large households have many dependents, such as elders and children, who are not able to make significant contributions to farm activities. It is similar with the family labour ratio. Households with more members of working age have reduced profit efficiency because they can use more home labour rather than hired labour in rice production.

The fragmentation of land was measured by the number of farm plots. A higher number of land plots may increase rice production costs (Tan *et al.*, 2010) or reduce production profit. This is a critical issue in the northern region of Vietnam, as discussed in the literature. According to Hung *et al.* (2007), a higher number of plots shows an increase in profit inefficiency or a decreased profit from rice production. This indicates that a greater number of plots can increase production costs for rice producers.

Table 15.5 Estimation for profit inefficiency

Variables	Para.	Coefficients	Standard errors	t-ratio
Constant	β_0	0.351***	0.172	2.04
Age of household head	β_1	0.001	0.001	1.19
Male household head	β_2	−0.061**	0.023	−2.67
Education of household head	β_3	−0.003	0.003	−0.94
Household size	β_4	0.060**	0.026	2.29
Family labour ratio	β_5	−0.042**	0.024	−1.75
Number of rice plots	β_5	0.052**	0.023	2.22
Monocropping	β_7	0.053	0.045	1.16
Row seeding technique	β_8	−0.060**	0.034	−1.74
Credit	β_9	−0.071	0.086	−0.83
Environmental factors:				
Soil quality	δ_1	0.074	0.146	0.51
Diseases	δ_2	−0.047**	0.023	−2.01
Irrigation	δ_3	−0.139**	0.057	−2.41
Water pollution	δ_4	0.043**	0.025	1.67
Total number of observations		**349**		

Notes: *** Significant at 1% level ($p < 0.01$); ** significant at 5% level ($p < 0.05$); * significant at 10% level ($p < 0.10$).

Monocrop cultivation in rice farming means that the household cultivates only one crop per year. This increases profit inefficiency. Reasons for monocrop cultivation may be that rice production is not profitable or that farmers can find other earning opportunities aside from producing rice. Using the row seeding technique in rice production can decrease cost and increase productivity. This is the technique that farmers in Vietnam's Red River Delta should use in rice farming when machinery is not employed for seeding. Additionally, diseases can decrease profit efficiency. Diseases in rice production are now common in Vietnam and in the Red River Delta in particular. However, farmers use a lot of pesticides and these can combat disease. Farmers are knowledgeable about the problems of disease and can take effective measures to remedy them. Water pollution is another serious problem in rice production. The pollution sources are from industries and handicraft production, in which untreated wastewater is discharged. In this study, rice producers in polluted regions made less profit because of greater expenditure on labour and the cost of other methods to deal with the effects of pollution, such as chemical fertilisers, organic fertilisers and pesticides.

Farmers can maximise profit efficiency by minimising profit inefficiencies. The factors significantly contributing to inefficiencies are household size, number of land plots and water pollution. Those that negatively contribute to inefficiencies are the gender of the household head, plant diseases and irrigation. The disease and irrigation variables are not well defined in the data. This is because 95% of farmers had encountered disease and 99% of the sample were in the region of the irrigation system (see Table 15.5). The reason for including these variables was to show the evidence that almost all farmers had encountered disease and also used irrigation in rice production.

Profit loss is defined as the amount that has been lost due to inefficiency in production given prices and fixed factor endowments and is calculated by multiplying maximum profit by – (1 – PE). Maximum profit per ha was computed by dividing the actual profit per ha of individual farms by the efficiency score (Rahman, 2003).

Table 15.6 shows the values of estimated profit loss per ha in different scenarios concerning key factors. These factors are household size, labour ratio, number of farm plots, number of crops cultivated per year, seeding

Table 15.6 Profit loss by key constraints

Farm-specific characteristics	*N*	Actual profit (VND 1000) per ha	Estimated profit loss (VND 1000) per ha	Profit efficiency
Profit loss by household size				
Small household	255	11,198.200	2758.276	0.764
Large household	94	10,191.030	3169.689	0.714
t-ratio		1.446*	–2.295**	2.991***
Profit loss by family labour ratio				
Low family labour ratio	171	9,470.631	2679.177	0.730
High family labour ratio	178	12,325.950	3051.527	0.771
t-ratio		–4.753***	–2.341**	–2.722***
Profit loss by number of farm plots				
1–3 farm plots	212	10,803.160	2562.956	0.766
> 3 farm plots	137	11,118.450	3342.805	0.729
t-ratio		–0.497	–4.915***	2.358**
Profit loss by monocropping				
Not monocropping	325	11,218.700	2885.818	0.744
Monocropping	24	6,975.811	2642.512	0.686
t-ratio		3.526***	0.769	1.803**
Profit loss by row seeding technique				
Not using row seeding technique	292	10,501.350	2927.031	0.740
Using row seeding technique	57	13,107.070	2572.244	0.810

(continued)

Table 15.6 *(Continued)*

Farm-specific characteristics	*N*	Actual profit (VND 1000) per ha	Estimated profit loss (VND 1000) per ha	Profit efficiency
t-ratio		−3.152***	1.643*	−3.484***
Environmental factor effects				
Farm has plant disease	2	10,225	918	0.900
Farm has no plant disease	347	10,930	2880	0.750
t-ratio		0.000	−1.850**	1.488*
Profit loss by soil quality				
Poor quality soil	140	9,560	2778	0.723
Fertile soil	209	11,842	2929	0.770
t-ratio		−3.000***	−1.000	−3.066***
Profit loss by irrigation				
Farms without irrigation	15	7009	3597	0.620
Farms with irrigation	334	11102	2836	0.757
t-ratio		−2.706***	2.000**	−3.73***
Profit loss by water pollution				
Farms without water pollution	170	12439	2959	0.776
Farms with water pollution	179	9490	2783	0.727
t-ratio		4.918***	1.099	3.271***
All farms	**349**	**10,926**	**2869**	**0.751**

Notes: Profit loss is computed from the maximum profit given prices and fixed factor endowments. Maximum profit per ha is computed by dividing the actual profit per ha of individual farms by the efficiency score.
* Significant at 10% ($p < 0.10$); ** significant at 5% ($p < 0.05$); *** significant at 1% ($p < 0.01$).

technique, plant disease, soil quality, irrigation and water pollution. In general, all farmers achieved around 75% profit efficiency. For each variable, we can see the eight difference values. They are the two values of observations (*N*), the two values of the actual profit, the two values of estimated profit and the two values of the profit efficiency.

For instance, the total number of small households was 255, and there were 94 large households. The profit that small households achieved per ha was VND 11,198,200 and for large households the figure was VND 10,191,030 per ha. However, small households show a lower profit loss (VND 2,785,276 per ha) than large households (VND 3,169,698 per ha). Additionally, small households reached higher profit efficiency (76.4%) compared to large households (71.4%).

This discussion has shed light on the factors that contribute to rice profitability. The knowledge in the literature shows that land fragmentation and environmental factors have significant effects on rice productivity (Hung *et al.*, 2007; Khai & Yabe, 2012; Long & Yabe, 2011; Rahman & Hasan, 2008). The research shows that environmental factors have an effect on rice profitability. Evidence has also been found in the literature (Khai & Yabe, 2013; Rahman & Hasan, 2011). In addition, this research found that rice farmers can maximise rice profit by better control of their production inputs, household resources and environmental factors.

6 Conclusion and Policy Implications

The estimation results show similar coefficients in the OLS and MLE approaches. Profit efficiency was around 75% on average and various environmental factors influenced this significantly, in addition to farm-specific characteristics. Plant disease, soil fertility, irrigation and water pollution were the four environmental factors that caused profit loss in rice production. The inclusion of environmental factors shows significant effects on profit efficiency. In other words, the environmental factors change the profit efficiency of rice production. Farmers can achieve around US$150 of profit per ha if they can control production constraints. Therefore, in order to maximise profit, rice farmers must overcome some obstacles in household characteristics such as labour and household size.

The policy implications are for the government to be involved in supporting the farmers to maximise their profit through methods such as land consolidation to minimise the number of farm plots and increase farm size, technology and water quality control. Advanced technology, such as row seed tools or machinery, should be applied and promoted in rice farming. These policy implications should be considered by the government in order to sustain rice production in the Red River Delta and across the country as well. Additionally, reducing household size is a recommendation that is in compliance with the family planning policy in Vietnam. Policymakers should be aware of the low labour productivity and high population growth rate of around 1% per year. Vietnam's population is about 96 million and ranked 15th globally (*Saigon Times*, 2019).

Finally, authorities are strongly advised to establish good governance of industrial environmental pollution by enforcement of environmental regulations and investing in infrastructure for water quality monitoring systems. This will firstly improve profit for farmers and, over time, help build the brand for Vietnamese rice for the benefit of the country.

Acknowledgements

We would like to express our gratitude to The Fuji Xerox Setsutaro Kobayashi Memorial Fund who granted us the necessary financial support to conduct the survey and research. Without this support, the research might not have been possible or successful. Next, we would like to give our thanks to the Rice Research Institute of Ha Noi University of Agriculture and its leaders for helping in the data collection process throughout the course of the research, and especially to the staff who provided data collection and administrative support. Finally, our warmest thanks go to Dr Huynh Viet Khai, Can Tho University, for his invaluable comments and proofreading of our final draft of the book chapter.

References

Aigner, D., Lovell, C.A.K. and Schmidt, P. (1977) Formulation and estimation of stochastic frontier production function models. *Journal of Econometrics* 6 (1), 21–37.

Ali, F., Parikh, A. and Shah, M. (1994) Measurement of profit efficiency using behavioural and stochastic frontier approaches. *Applied Economics* 26 (2), 181–188.

Ali, M. and Flinn, J.C. (1989) Profit efficiency among basmati rice producers in Pakistan Punjab. *American Journal of Agricultural Economics* 71 (2), 303–310.

Dung, N.M. (2009) Compliance of paper-making plants with environmental regulation in Bac Ninh Province, Vietnam. EEPSEA Research Report No. rr2009061.

Farrell, M.J. (1957) The measurement of productive efficiency. *Journal of the Royal Statistical Society. Series A (General)* 120 (3), 253–290.

Hung, P.V., MacAulay, T.G. and Marsh, S.P. (2007) The economics of land fragmentation in the north of Vietnam. *Australian Journal of Agricultural and Resource Economics* 51 (2), 195–211.

Hyuha, T., Bashaasha, B., Nkonya, E. and Kraybill, D. (2007) Analysis of profit inefficiency in rice production in Eastern and Northern Uganda. *African Crop Science Journal* 15 (4), 243–253.

Khai, H.V. and Yabe, M. (2011) Technical efficiency analysis of rice production in Vietnam. *Journal of ISSAAS* 17 (1), 135–146.

Khai, H. and Yabe, M. (2012) Rice yield loss due to industrial pollution in Vietnam. *Journal of US-China Public Administration* 9, 248–256.

Khai, H. and Yabe, M. (2013) Impact of industrial water pollution on rice production in Vietnam. In N.W.T. Quinn (ed.) *International Perspectives on Water Quality Management and Pollutant Control* (pp. 61–85). IntechOpen.

Kolawole, O. (2006) Determinants of profit efficiency among small scale rice farmers in Nigeria: A profit function approach. *Research Journal of Applied Sciences* 1 (1), 116–122.

Long, H.V. and Yabe, M. (2011) The impact of environmental factors on the productivity and efficiency of rice production: A study in Vietnam's Red River Delta. *European Journal of Social Sciences* 26 (2), 218–230.

Mahanty, S. and Dang, T.D. (2013) Crafting sustainability? The potential and limits of institutional design in managing water pollution from Vietnam's craft villages. *Society and Natural Resources* 26 (6), 717–732.

Meeusen, W. and Van Den Broeck, J. (1977) Efficiency estimation from Cobb–Douglas production functions with composed error. *International Economic Review* 18 (2), 435–444.

MOLISA (2009) *More than 90% of Trade Villages across the Country Violate the Legislation on Environment and Labor Safety* (translated). See http://www.molisa.gov.vn/Pages/tintuc/chitiet.aspx?TinTucID=11758&page=36 (accessed 20 February 2020).

Quynh, N.K., The, T.V. and Luu, L.V. (2003) Assessing participatory rural environmental management in the craft villages. Report submitted to FAO. Vinh City: Vietnam Agriculture Sciences Institute (VASS).

Rahman, S. (2003) Profit efficiency among Bangladeshi rice farmers. *Food Policy* 28 (5–6), 487–503.

Rahman, S. and Hasan, M.K. (2008) Impact of environmental production conditions on productivity and efficiency: A case study of wheat farmers in Bangladesh. *Journal of Environmental Management* 88 (4), 1495–1504.

Rahman, S. and Hasan, M.K. (2011) Environmental constraints and profitability relationships in agriculture: A case study of wheat farming in Bangladesh. *Journal of the Asia Pacific Economy* 16 (4), 630–643.

Saigon Times (2019) Vietnam's population reaches 96 million, ranks 15th globally. *Saigon Times*, 7 November. See https://en.thesaigontimes.vn/tinbaichitiet/69738/ (accessed 11 October 2019).

Sherlund, S.M., Barrett, C.B. and Adesina, A.A. (2002) Smallholder technical efficiency controlling for environmental production conditions. *Journal of Development Economics* 69 (1), 85–101.

Sidhu, S.S. and Baanante, C.A. (1981) Estimating farm-level input demand and wheat supply in the Indian Punjab using a translog profit function. *American Journal of Agricultural Economics* 63 (2), 237–246.

Tan, S., Heerink, N., Kuyvenhoven, A. and Qu, F. (2010) Impact of land fragmentation on rice producers' technical efficiency in South-East China. *NJAS – Wageningen Journal of Life Sciences* 57 (2), 117–123.

Vietnam Food Association (2018) *Vietnam Rice Exports in the Period of 1989–2017*. See https://www.pinterest.com/pin/701294973205302783/ (accessed 2021).

Vu, L.H. (2008) Essays on the economics of food production and consumption in Vietnam. PhD dissertation, University of Minnesota.

WB (2008) *World Development Report: Agriculture for Development*. Washington, DC: The World Bank.

Conclusion: Development – Prosperity or Disparity?

Pahlaj Moolio and Michal Apollo

'May you live in interesting times.' The oft-quoted expression, seemingly a benediction, is usually said in an ironic way. Life is much better in the 'uninteresting times' of peace and tranquillity than in 'interesting' ones, which are usually times of trouble. And here we are, living in the most interesting times in the history of humanity – in the times of extremes of practically everything, be it poverty or prosperity or the heights of scientific discoveries or the total inability to fulfil basic human needs.

The intent of this volume was to bring together interdisciplinary and multidisciplinary perspectives to broaden the understanding of the concepts of poverty and development. The word 'poverty' means the condition of being poor. The overall aim of economic development has been to eradicate poverty. However, there is widespread concern that economic development has not been shared fairly within and among nations, and the gap between the rich and the poor has been widening, causing extreme inequality.

This situation has caused unprecedented suffering in society, especially during the last ten years since the global financial crisis and the Great Recession that shook the entire world, causing extreme hardship, especially for the poorer people in every land. During this period, the fortunes of the rich, predominantly men, have risen dramatically, resulting in a doubling of the number of billionaires, with a new billionaire added every three to four days in 2017 and 2018. At the same time, almost half of humanity is compelled to live on less than US$5.50 a day while the world's richest 1% has more than twice as much wealth as 6.90 billion people combined. There are more billionaires worldwide now than ever before and, together, they have more wealth than the total accumulated by 60% of the planet's population. In the meantime, hundreds of millions of people are still living in extreme poverty. This is an example of extreme inequality of wealth in an increasingly interconnected and interdependent digitalised world.

Sadly, many governments, in both the East and the West, have fuelled and continue to fuel this crisis of inequality. They are under-taxing corporations and wealthy individuals, but underfunding vital public services like healthcare, education and poverty eradication schemes. For instance, only 4% of every dollar of tax revenue comes from taxes on wealth, as the super-rich avoid as much as 30% of their tax liability. This financial anomaly has hit the poor the hardest, resulting in huge human costs, with women and girls being marginalised and indigenous people suffering the most. This inequality is keeping people trapped in poverty or pushing them further towards extreme poverty, causing suffering for humanity's collective life because the welfare of any segment of humanity is tied up with the welfare of the whole.

We are living in a digital age, facing various challenges, and our minds are in a state of utter confusion. Undoubtedly, the world is going through great turmoil – a rapid digital transformation exposed by a rising tide of information, causing disorder and confusion. It is a chaotic phase of transition through which humanity is collectively confronted with the difficult reality of unprecedented global uncertainty and forces of fragmentation and integration operating side by side. With the use of digital technology, unacceptable amounts of wealth are being amassed by some, and this situation is further aggravated by the uneven distribution of wealth between the nations and within nations.

Mother Earth itself is now telling us through physical signs of climate change (i.e. increasing land and ocean heat, accelerating sea level rise and melting snow) that overconsumption of natural resources, uncontrolled urbanisation in the name of modern civilisation, climate change induced migration, unequal access to ecosystem services and the present and potential environmental, societal and economic conflicts brewing in tourist regions are leading humanity to disaster. Contemporary tourism, although known worldwide as a significant catalyst for economic growth and job creation, causes various kinds of costs to the economy, society and the environment. On the other hand, agritourism, ecotourism and slow tourism have great potential for creating sustainable rural development with specific effects on the environment, agricultural heritage or economic growth, and contribute to reducing poverty through regular income distribution and low-carbon travel and accommodation. Nevertheless, the slogan of sustainable development, so popular in planning conferences and development documents, is not implemented in practice at all.

While the incoherence within the current economic and social systems is seemingly leading to their collapse, clearing away some obstacles to new sustainable systems, humanity is presently faced with the headwinds of a stubborn resistance to change. The youth are protesting in the streets as they know very well that their future is in danger. This is because rules, structures and systems, designed by the rich for the rich, have manifestly failed to serve the interests of all people.

This rapid transition and its various challenges represent a compelling invitation for all of us to act not only wisely but quickly as well, strengthening the forces of integration and solidarity to create coherence between sustainable economic, social and environmental systems. These systems will be able to lead the transformation towards a more sustainable and equitable world that draws on the abundant renewable energy sources of this planet. This calls for a fundamental reorganisation of the economic affairs of human society that can advance the collective prosperity of human beings and leave no-one behind. Founded on the principles of social justice and generosity, collaboration and mutual assistance, this reorganisation would make an inherent moral dimension the principle for the generation, distribution and utilisation of wealth and the world's resources. Wealth must serve humanity.

There are many caring people in the world who aspire to improve the conditions of human society, yet most of them have neither a clear concept of the sort of world they wish to build nor any idea of how to go about it. Thus, they are reduced to combating every apparent evil that catches their attention. It is vital, therefore, to understand the underlying conditions that lead to fragile and unsustainable systems, and to identify new underlying conditions – the best foundations to help humanity build a prosperous, just and more sustainable civilisation. For example, the current measures of success, such as GDP growth, do not take into account sustainability or the well-being of people and planet. New indicators of success should be introduced in the business world and economic systems. Definitions of success are usually framed as serving the apparent interests of each nation, corporation, local community or individual, but now we should rather consider how we can build a more realistic global consciousness of interdependence.

The urgency of the current global situation demands that each one of us push forward innovative ideas that were thought impossible or too audacious in the past. How do we find the right questions and actions to allow each of us and our organisations to address excessive carbon emissions, the unprecedented loss of biodiversity and all the consequences we are already witnessing from climate change? While we do not have all the answers, we must now work to reinvent the economic and social systems and the very purpose of business, striving for the common good and not just for the few at the top at a cost to the many at the bottom.

But where does one start to help this transition from the current negativity, polarisation and unsustainable trends to more prosperous, just and sustainable organisations of economic and social systems? What humanity has been witnessing over the past few decades of human history is the importance of a few parallel actions.

The most important aspect of this is our understanding of human nature. For example, unquestioning agreement is given to the proposition that human beings are inherently selfish and aggressive. They are therefore incapable of erecting a social system at once progressive and peaceful, dynamic and harmonious. We should rather be fostering an understanding of the true nature of humanity. We should strive to perceive the nobility in every human being – rich or poor, man or woman, old or young, city dweller or villager, employee or employer – irrespective of ethnicity or religion, caste or creed. We should attend to the needs of young people and foster in them confidence in the future so that they may prepare themselves adequately for service to humanity.

There is a need to engage each and every individual in designing the ideal sustainable social and economic systems – the scenario of the future to which humanity wishes to transition. Thus, creating a vision is vital. All businesses and economic and social systems should focus on putting a human purpose at the heart of the economy, providing the material means to enable every human being to prosper and develop rationally, not just to consume in a consumer culture. The economic system should be altruistic and cooperative, with businesses at a human scale.

We should develop our and our colleagues' capacities to engage in key social discourses and to create unity in every aspect of life. We should learn to foster environments of trust where diverse stakeholders have the volition to come together, co-design a vision and create a shared understanding of the underlying assumptions that are key to unlocking the sustainable economic and social future that we urgently need to implement.

The first step for each of us should be to create a new state of mind – one in which a clear aim is set, founded on strong underlying values. Second, we need to apply this vision in our actions to explore the way forward. Third, we must follow our actions with a period of shared reflection upon the results. This reflection would not be a waste of time, but a key moment to come together and understand and reconnect our actions with the purposes that will benefit the general population of the world.

And here we are, at the cusp of a new era. We can be the moral people building the future and we can walk this shared path:

> The pathway to sustainability will be one of empowerment, collaboration and continual processes of questioning, learning and action in all regions of the world. It will be shaped by the experiences of women, men, children, the rich, the poor, the governors and the governed as each one is enabled to play their rightful role in the construction of a sustainable society. As the sweeping tides of consumerism, unfettered consumption, extreme poverty and marginalization recede, they will reveal the human capacities for justice, reciprocity and happiness. (BIC, 2010)

The only question is, will we choose this path?

Reference

BIC (2010) Rethinking prosperity: Fostering alternatives to a culture of consumerism. *Bahá'í International Community*, 3 May. See https://www.bic.org/statements/rethinking-prosperity-forging-alternatives-culture-consumerism (accessed 10 May 2020).

Index

Lightning Source UK Ltd.
Milton Keynes UK
UKHW022023051021
391659UK00004B/268

9 781845 418465